M 691

OXFORD MEDICAL PUBL

GW00631154

Accidents and emergenc

Accidents and emergencies

A PRACTICAL HANDBOOK
FOR PERSONAL USE

R. H. HARDY
Formerly Consultant in Accident and Emergency Medicine
Hereford General Hospital

FOURTH EDITION

OXFORD
OXFORD UNIVERSITY PRESS
NEW YORK   TORONTO
1985

Oxford University Press, Walton Street, Oxford OX2 6DP

London New York Toronto
Delhi Bombay Calcutta Madras Karachi
Kuala Lumpur Singapore Hong Kong Tokyo
Nairobi Dar es Salaam Cape Town
Melbourne Auckland

and associated companies in
Beirut Berlin Ibadan Mexico City Nicosia

Oxford is a trade mark of Oxford University Press

© Robert Dugdale and R. H. Hardy 1976, 1978, 1981, 1985

First edition published 1976
by Robert Dugdale, c/o Corpus Christi College, Oxford OX1 4JF
Reprinted 1976 (with corrections), 1977
Second edition published 1978 by Oxford University Press
Reprinted with corrections 1978   20-3-85
Third edition 1981. Fourth edition 1985
                               168345   m 691

British Library Cataloguing in Publication Data

Hardy, R. H.
Accidents and emergencies: a practical
handbook for personal use. — 4th ed. —
(Oxford medical publications)
1. Emergency medicine
I. Title
616'.025   RC86.7
ISBN 0-19-261527-0

Library of Congress Cataloging in Publication Data

Hardy, R. H.
Accidents and emergencies
(Oxford medical publications)
Bibliography: p.   Includes index.
1. Emergency medicine — Handbooks, manuals, etc.
2. Accidents — Handbooks, manuals, etc.   I. Title.
II. Series.   [DNLM: 1. Accidents — handbooks.
2. Emergencies — handbooks.   3. First Aid — handbooks.
WA 39 H272a]
RC86.7H37 1984   616'.025   84-20738
ISBN 0-19-261527-0 (pbk.)

Printed in Great Britain
at the University Press, Oxford
by David Stanford
Printer to the University

Acknowledgements

Author and publisher would like to thank the following for permission to use material reproduced in this book: Smith and Nephew Pharmaceuticals Ltd for Figs 4 and 5; B. H. Bass MD, FRCP, of the Good Hope Hospital, Sutton Coldfield, for the circular on gas gangrene; Brian Thomas, FRCS, for permission to illustrate his superb radial palsy splint.

The author's thanks in particular go to Dr J. O. P. Edgcumbe for insisting on the article on **Haematological emergencies** and to Dr Jeff Kramer for his help in preparing it; to Dr I. R. Ferguson for checking the bacteriology of the PUO table; finally to Mrs Brenda Prosser for turning illegible manuscript into rational type.

Preface to the fourth edition

Many entries have been brought up to date, many errors cor-
rected, and gaps and omissions filled; obscurities have been clari-
fied and illustrations added or improved. This has largely been
achieved by using the blank pages, in an effort to keep the book's
size and price suited to the pockets of hospital staff as well as the
bags of general practitioners. There are still plenty of blanks for
the user to fill.

The book still has many shortcomings, of which the author is
acutely aware, but its usefulness has been, he hopes, increased. As
the Accident and Emergency revolution begins to crystallize into
set forms, set patterns replace open options. As the patterns set,
disagreements deepen and fashions of thought become more and
more influential. The book has tried to combine common sense
with an open mind and it is hoped that the new index will make
both more easily accessible. The bibliography reflects the frequency of new and excellent publications in our field and the
need for wide reading and exchange of ideas if we are to meet our
obligations to our patients.

The basic layout is still alphabetical by subjects, but the contents all appear in one index — the main headings in **BOLD CAPI-
TALS**, subheadings in **bold type**, and other entries in plain type.

RHH

Preface to the first edition

I have found a continual need of a handbook to give to medical
and nursing staff coming to accident and emergency work for the
first time, and I know that many others in my place have found
the same.

This one is based on the handbook which grew up in Hereford,
but it has been largely rewritten after the critical scrutiny of
many friends and colleagues, in particular Mr James Scott, FRCS,
who has gone through the text with a thoroughness that I cannot

thank him enough for, and Miss Sheila Christian, FRCS, who has made many useful suggestions and criticisms.

Any textbook of accidents and emergencies will have limitations because of the lack of any generally accepted practice in the field, the absence of any received dogma, and the fact that all medical and surgical practice is in a state of flux. Besides which, the variations between different parts of the country and different hospitals make any definitive textbook of little value except as a basic source book. So the solution is offered here of a skeleton text upon which the casualty officer can build his own code with the help of his seniors in the special fields of accidents and emergencies, orthopaedics, and all the other specialities which overlap with them.

The opinions and guidance offered are often heterodox and sometimes frankly contentious in the hope that they may stimulate radical rethinking of current practice and a revaluation of every user's concepts in the light of his own experience as well as ours. The text on the left-hand page is only a framework upon which the reader can build. What will be really valuable is what is written on the right.

Accident and emergency work is emerging as a career which attracts more medical talent each year because of the increasing realization of its capacity for growth and improvement in giving help to the injured and acutely ill, and the endless opportunity and interest it offers to its practitioners as its scope and skill develops. This very tentative compilation is made in the hope that beginners of all sorts may find it a useful basis for building up their own expertise.

A casualty officer in the medical organism has been compared to the hand in the individual's constitution — both are the most highly unspecialized organs in the body. So the keynote of a book designed to help the first to be fully effective has to be adaptable versatility.

My thanks to my publisher, Robert Dugdale, are limitless for his faith in my undertaking and for the endless pains he has taken in trying to make a silk purse out of a sow's ear. I have been continually helped and encouraged by his critical enthusiasm and hope that he will have no occasion to regret his daring.

Needless to say the responsibility for everything written here is my own, but it would unquestionably have been far worse without the help of all the many people who have been so generous with their aid.

RHH

Abbreviations

A & E	accident and emergency
BMJ	*British Medical Journal*
BP	British Pharmacopoeia
°C	centigrade
CAT	computerized axial tomography
cf.	*confer* (Latin for 'compare')
cm	centimetre(s)
CNS	central nervous system
CVP	central venous pressure
d.c.	direct current
DIC	disseminated intracapillary coagulation
DVT	deep-vein thrombosis
EB	Epstein—Barr
ECG	electrocardiogram
ECM	external cardiac massage
e.g.	*exempli gratia* (Latin for 'for example')
ENT	ear, nose, and throat
etc.	*etcetera* (Latin for 'and the rest')
°F	degrees Fahrenheit
g	gram(s)
GA	general anaesthetic
GLC	gas liquid chromatography
GP	General Practitioner (of medicine)
H_2O_2	hydrogen peroxide
ICI	Imperial Chemical Industries
ICU	intensive care unit
i.e.	*id est* (Latin for 'that is')
IM	intramuscular(ly)
IMI	intramuscular injection
in	inch(es)
inj.	injection
Ig	Immunoglobulin
ITU	intensive therapy unit
IV	intravenous(ly)
IVI	intravenous injection

J	Joule(s)
JVF	jugular venous filling
JVP	jugular venous pulse
kg	kilogram(s)
kNm^{-2}	kilonewton(s) per square metre
kPa	kilopascals
l	litre(s)
lab.	laboratory
lb in^{-2}	pound(s) per square inch
LSD	lysergic acid
mega	× 1 million
mEq.	milli-equivalents
mg	milligram(s)
μg	microgram(s)
μg/ml^{-1}	microgram(s) per millilitre
ml	millilitre(s)
mm	millimetre(s)
mmol	millimole(s)
NB	*nota bene* (Latin for 'note well')
NSAID	non-steroidal anti-inflammatory drug
OM	occipito-mental (diameter of the skull)
ORIF	open reduction and internal fixation
path.	pathological
PID	prolapsed intervertebral disc
POP	plaster of Paris
PPV	positive-pressure ventilation
PUO	pyrexia(s) of unknown origin
q.v.	*quod vide* (Latin for 'which see')
RTA	road traffic accident(s)
SOS	*si opus sit* (Latin for 'if there is need')
strep.	streptococcus
T	temperature
t.d.s.	*ter die sumendum* (Latin for 'to be taken thrice daily')
U	units
UK	United Kingdom
USA	United States of America
USNF	United States National Formulary
VD	venereal disease
w/v	weight per volume

ABDOMINAL INJURIES

Common in RTA, horse-riding accidents, gun-shot wounds, stabbings, agricultural accidents, and even sometimes in domestic accidents.

Blunt abdominal trauma

A solitary injury due to (e.g.) steering wheel or lap-strap compression is easy to spot: tenderness, guarding and rigidity are diagnostic. Bruising of the soft abdominal wall or fabric imprints on to the skin over soft parts are serious signposts. Associated low rib-fractures alert the examiner to the possibility of splenic or hepatic injury.

If an abdominal injury is one of many, especially if consciousness is lost, it is very easy to miss it.

In all cases of multiple injuries where there is any risk of abdominal injury as well, girth measurements should be made at a marked level every ten minutes and recorded. Increasing girth is a warning of bleeding or ileus.

Ruptured spleens may bleed slowly; they may also develop a subcapsular haematoma which can rupture in 7—15 days.

Ruptured liver is a common event in blunt abdominal trauma and is often only clearly identified at laparotomy, which has been decided on in response to a patient's failure to stabilize his pulse and blood pressure after injury.

Traumatic perforation of the bowel usually occurs at a site where external compression meets internal resistance — e.g. where the duodeno-jejunal junction rests on the 3rd lumbar vertebra. Tears of the meso-colon and other mesenteries in the midline can also occur.

Renal damage may arise from direct violence to the loin; pelvic viscera can be involved in injuries to the bony pelvis, notably the bladder and the female urethra. The male urethra is injured less frequently and the ureter very rarely (perhaps only in cases of fracture-dislocation of the sacro-iliac joint).

Sharp abdominal trauma (cf. Stab wounds)

In general, early operation is not all that important in stab wounds, but awareness of the possibility of multiple perforations and the

Sharpened bicycle spokes pushed into the abdomen can give multiple bowel injuries with minimum external marking.

likelihood of bowel trauma in the absence of a clear wound-track through the abdominal parietes will make the observer alert to changes in the patient's physical signs. He must share this awareness with nursing staff responsible for trolley-side observations.

Aortic or common iliac lacerations due to blunt or sharp trauma rarely cross the therapeutic threshold.

In identifying intraperitoneal bleeding, needle-endoscopy can be used in experienced hands under local anaesthesia, but peritoneal lavage is the most reliable in A & E units: use a dialysis catheter introduced through a small incision, suprapubically in the mid-line, half-way between pubes and umbilicus. Warm normal saline is used for lavage and a 2-way tap for drainage into a bottle with under-water seal.`

ABORTIONS (cf. Obstetric emergencies)

Threatened, inevitable, incomplete, or complete, abortions are unpredictable and often combined with anxiety and fear. They are best dealt with in a specialist unit, and should be given the briefest of examinations in the ambulance or car and transmitted as soon as possible to their proper destination, unless urgent resuscitation is required on arrival.

ABSCESSES (cf. Hand infections)

Small ones can often be incised under intradermal local anaesthetic, but it is difficult to ensure full breakdown of loculi and proper clearance in big ones. Infected sebaceous cysts respond well. Some authorities dislike local infiltration of abnormal skin but no ill effects have been observed in practice. General anaesthetic is preferred if available.

Other abscesses are best opened under general anaesthetic or IV narcanalgesia with doxapram safeguard (q.v.). If of any size they are traditionally fully explored by the gloved finger and rubber/plastic drain, or antiseptic tulle gras or ribbon-gauze inserted to ensure proper drainage. Drainage is prevented by tight packing. This method should now be relegated to history.

Combined drainage and systemic antibiotics give quicker healing than either separately, and should always be used together.

A high index of suspicion in casualty officers is of far more value than special investigations if abdominal injuries are not to be missed.

IVI Syntometrine may control bleeding. Septic abortion should receive IVI clindamycin 300 mg as well.

If you don't like the look of it refer it to a senior member of the staff.

Subcuticular abscesses should be *completely* unroofed.

Test the urine of *every* case for sugar.

Without doubt the recent development of Ellis's method of treatment of major abscesses supersedes all the traditional ones (*The casualty officer's handbook*, 4th edn: see **Bibliography**, p. 169).

There are 8 steps:

1 Injection of the right antibiotic(s) ½–1 hour pre-operatively
2 Anaesthesia
3 Incision of sufficient size to admit one exploring index finger
4 Digital exploration and evacuation of pus
5 Curettage of the entire pyogenic membrane and dry gauze toilet of the cavity
6 Irrigation, without raising the pressure inside the cavity, using hydrogen peroxide solution (20 ml syringe and filling 'quill')
7 Sutures so placed as to occlude the cavity
8 Antibiotic continued orally. After the first 24 hours twice daily hot baths are comforting and helpful.

The choice of antibiotic is important: all abscesses below the waist merit treatment with clindamycin, as infections are usually mixed and include anaerobes (300 mg by IM injection pre-operatively — 150 mg orally, 6-hourly, for 5 days). An increasing number of other abscesses contain anaerobes as well and the characteristic stink of the pus generally gives them away. If the infection is clearly staphylococcal, flucloxacillin is given by injection and by mouth. If there is a possibility of a mixed infection, penicillin-G 600 mg + flucloxacillin 500 mg for pre-op. injection and Augmentin 750 mg 8-hourly orally to follow.

Many path. labs provide a *GLC profile* which provides evidence of anaerobic infection within the hour.

The benefit to patients of this out-patient treatment of major abscesses is incalculable: 2 visits (for operation and removal of sutures) instead of dozens; minimal pain and inconvenience, brief recovery period, diminished recurrence. It does a great deal, too, to relieve the pressure of repeated attendances at A & E departments. Traditionalists sometimes cavil, but only those who haven't tried it.

Don't forget: it depends for its success on thorough and

Topical povidone ointment after bathing is antiseptic and soothing.

The legitimate successors to Ellis's method have advocated single-dose antibiotic only (Blick and others, *BMJ* **ii** 111 (1980)).

complete surgery and the appropriate use of antibiotics; it is *not* suitable for hand infections; it is suitable for *all* other major abscesses.

Pilonidal abscess

The important feature of this mid-line sacro-coccygeal abscess is its central sinus extending upwards for varying distances (up to 65 mm; 3½ in) from its orifice. The sinus is epithelialized and may contain hairs. It should be opened on to a probe inserted to the full extent of the sinus and treated by the clindamycin method described, after dissection and energetic curettage of the epithelial tract and all its ramifications.

ACCIDENTS, MAJOR

With head, or multiple and head, injuries the routine treatment is simple and must always be followed in this order:

1   Ensure **airway**, with cuffed tube if necessary, and ventilation — get an anaesthetist fast if in difficulty or doubt.
2   Stop major external **bleeding**, usually by local pressure.
3   Strip and **examine**.
4   Step up **IV fluids** if necessary — 4 may need to precede 3 — bilaterally if required and always in the arms if possible, having first taken blood specimens for cross matching and any relevant baseline estimations.
5   Investigate **injuries**, and record their nature and significance.
6   **Treat** if required.
7   The importance of getting as clear a **history** as possible must not be forgotten, but circumstances may prevent it.
8   Single shot lateral X-ray of neck in all unconscious patients.

Where accident and intensive care units are closely associated, a CVP catheter may be inserted early, but for the majority of accident units it is inappropriate. Where this is to be done the safest and best method, and often the quickest, is by dissection of the subclavian vein at its junction with the axillary vein — i.e. where the clavicle crosses the first rib. The CVP catheter is inserted after digital occlusion of the subclavian vein and the venous incision is carefully closed by superficial interrupted 5/0 Prolene stitches. The skin is similarly closed and carefully dressed. The

Otherwise the Ellis (or Leeds) method of abscess surgery is followed.

Use clear plastic tubes in major injuries who may need ventilation.

In cases of major or multiple injuries give steroids (p. 5a).

skin wound is sealed with povidone-iodine ointment (Betadine). Full surgical aseptic precautions are observed throughout.

Any major accident demands a combined operation and the more relevant help you get in, after the first assessment of the situation, the better for everyone — especially the patient. Evidence from witnesses and ambulance personnel can be of vital importance in the first assessment.

With practice you should be able to make a reliable assessment of the number and severity of injuries in a 5-minute examination. Your findings should be briefly, clearly, and definitely recorded, starting at the top, like this:

Part of the body	*Injuries*
Head and oropharynx	. . . . .
CNS	. . . . .
Chest	. . . . .
Abdomen	. . . . .
Pelvis	. . . . .
Upper limbs	. . . . .
Lower limbs	. . . . .
Skin	. . . . .
Other	. . . . .

Keep this list of injuries separate from observations of the patient's condition, especially surgical shock and circulatory failure.

ACCIDENTS, ROAD TRAFFIC

Injuries are often multiple. It is easy to overlook an associated injury by concentrating on the primary one, e.g.:
1 In cases of dashboard injury to the knee it is important to X-ray the ipsilateral hip to exclude fracture/dislocation.
2 In cases of head injury with loss of consciousness it is easy to miss an associated fracture/dislocation of the cervical spine.
3 Severely injured people are often also drunk (and vice versa) — ask for an urgent blood-alcohol (if available) to help clarify the severity of their head injuries.

You will quickly adopt your own or your department's techniques for the reception and treatment of such patients. It will not be

Steroids

All major and multiple injuries should be protected against:
   Fat embolism
   Shock-lung
   Disseminated intravascular coagulation
   Post-traumatic capillary permeability
The first and most vital remedy is blood-volume replacement, but the role of steroids must not be forgotten. Add 2 g of methyl-prednisolone to the first unit of blood or plasma substitute. This may be preventative and probably is. Cf. **Chest injuries**, p. 34.

One of the few bits of gadgetry which is of real value is an auto-matic, recording, vital signs monitor. It gives readings of pulse and blood-pressure at set intervals and a print-out to send to the ward with the patient. An alarm (pre-set) warns of undue changes.

helpful to write out a scheme here. Make your own, and make it infallible, and record it on the opposite page.

ACTINOMYCOSIS

A rare cause of subacute submandibular abscess, likely to be secondary to periodontal infection. Advanced actinomycosis is unheard of today in medically sophisticated countries. Treatment is by surgical drainage and IM penicillin.

ADHESIVES

Cyano-acrylate adhesives have introduced a new dimension into emergency production. Their adhesion is instant. A child can stick fingers indissolubly together, gum up his eyes, nose, and mouth with a single smearing gesture and die of suffocation. The manufacturers comfortingly advise peeling apart adherent skin surfaces after applying warm soap and water. Emergency laryngotomy may however be required.

Every accident department should obtain the manufacturers' well-prepared advice-sheet, available on application to Loctite (UK) Ltd, Welwyn Garden City, Herts, AL7 1JB.

ANAESTHESIA, LOCAL AND REGIONAL

Digital nerve block for fingers and toes

If this is done well, it works with 100 per cent success. If it is done badly it is ghastly and saps the patient's confidence like nothing else.

Where appropriate, the injection should be given into the interdigital web — lignocaine 1% 3—5 ml each side in adults. The needle should point towards the digital nerve (see Fig. 1). If there is no web, local anaesthetic should be injected from the dorsum of the finger until the bulge is palpated in the volar aspect (about 2.5 ml in children, 3—5 ml in adults). You *must* wait at least 5 minutes. Then, if all is not well, repeat the dose. A tourniquet can be applied after injection if prolonged operation is anticipated. Otherwise it is superfluous and may be painful. The injection

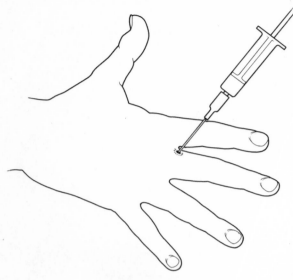

Fig. 1.

must be given into healthy tissue. In the presence of infection (e.g. a large paronychia) larger quantities are generally required. Thus it is wise to give more to the nerve on the affected side of the digit.

Intermetacarpal block

should always be given from the dorsum of the hand and it is wise to give 8–10 ml of lignocaine 1% either side of the relevant metacarpal. A tourniquet to the forearm (sphygmomanometer cuff is safest) will increase its efficacy.

Wrist block (see Fig. 2)

Palpate the pisiform and hamate bones on the volar surface of the wrist medially, and the tubercule of the scaphoid laterally. Anaesthetize the skin. Inject lignocaine 1% 5 ml immediately lateral to the pisiform and superficial to the flexor retinaculum at the level of the distal skin-crease. Then inject lignocaine 1% 2 ml subcutaneously immediately medial to the tubercule of the

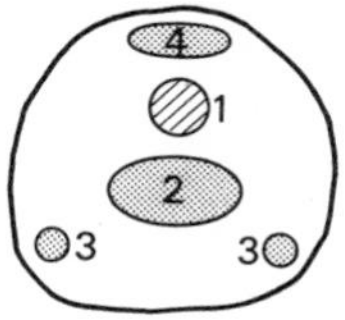

Fig. 1a. Cross-sectional schema of finger showing position of the digital nerves. (1) Phalanx. (2) Flexor tendon and fibrous flexor sheath. (3) Neuro-vascular bundle. (4) Extensor tendon.

Do not forget 1 ml of local anaesthetic into the dorsum of toe or finger for branches of a variable dorsal nerve supply. *Never use adrenaline in fingers*.

Marcaine 0.5% gives prolonged anaesthesia.

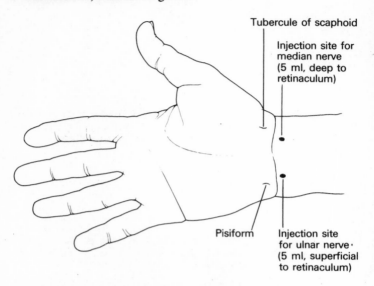

scaphoid. Then press the point of the needle through the flexor retinaculum and inject 5 ml. This may need considerable force as the space in the carpal tunnel is restricted. A cuff can be used as above.

In 15 minutes anaesthesia of the palm and fingers should be complete. Median paraesthesiae give a good indication that the needle is in the right layer. If they are obtained, a fractional withdrawal is indicated to avoid injection into the substance of the nerve itself. You can generally feel the needle pierce the flexor retinaculum with a pop, similar to that noticed as the lumbar puncture needle enters the vertebral canal.

You must put 5—10 ml of lignocaine under the skin on the dorsum of the wrist to catch the cutaneous branches of the radial nerve.

Brachial plexus block

This is less easy and best left to the professionals. It can be complicated by a subclavian haematoma or traumatic pneumothorax.

Fig. 2 (opposite). Tendon landmarks: between palmaris longus and flexor carpi radialis for the median and between palmaris longus and flexor carpi ulnaris for the ulnar nerve.

9 Anaesthesia, local and regional

Axillary block

Safer than brachial plexus block, but the time-consuming and not always successful. It is described in standard textbooks but requires detailed illustration not practicable in a brief handbook. See **Bibliography**.

Intravenous regional block (Bier's block)

For use in the elderly in reducing e.g. Colles's fractures (see **Fractures**). Uncomfortable, time-consuming, and not always successful. I find that local infiltration of the haematoma or IV narcanalgesia (q.v.) seems to give better results, and is kinder.

Other regional anaesthetic blocks

These require practice and skill. They are best learnt by demonstration. See **Bibliography** and p. 64.

Intradermal local anaesthesia

See Abscesses.

Subcutaneous anaesthesia

Routinely used for minor operations and skin suturing. Do not proceed until you have verified the procedure's effectiveness by testing for absence of pin-prick sensation.

ANAPHYLAXIS, ACUTE

Strictly a hypersensitivity reaction to foreign protein, characterized by an initial sensitizing dose and an outpouring of histamine or histamine-like substances in response to a second effective dose of the original specific allergogen. It can be tissue specific (e.g. dermal in acute allergic urticaria) or general (e.g. loss of consciousness, asthma, urticaria, cardiac arrhythmia, and peripheral vascular collapse following a wasp sting in a sensitized subject). The term is also loosely used to describe anaphylactoid response to single-dose assaults from such dramatic histaminagogues as snake or scorpion bites, jelly-fish and insect stings.

Hypersensitive subjects who may need instant treatment for

Femoral nerve block

Should be done on reception of every patient with fractured shaft
of femur; it gives rapid pain-relief and allows skin-traction splintage with comfort and ease. Stand on the contralateral side facing
the patient's head; put, and keep, the non-dominant index finger
on the femoral artery where it crosses the pelvic brim; inject 10 ml
of 0.5% bupivacaine, with adrenaline 1/200,000, fanwise and
laterally to the fixed index. Wait 10 minutes and proceed with
your manipulations gently, firmly, and swiftly.

Upper limits

Suggested totals in order to avoid undesirable side-effects of local
infiltration:

    Lignocaine 1% plain — 20 ml (200 mg)
    Lignocaine 1% with adrenaline 1/80,000 — 30 ml (300 mg)
    Bupivacaine 0.5% plain — 30 ml (150 mg)
    Bupivacaine 0.5% with adrenaline 1/200,000 — 45 ml (225 mg)

life-threatening reactions out of range of immediate medical care should always ·carry an adrenaline nebulizer (e.g. Medihaler epi — Riker), which can be life-saving.

In A & E units the nebulizer can be used as a first-aid measure but should be followed by IV adrenaline 0.1%, 0.1 ml each minute (or 1 ml of 0.01% each minute) titrated against the control of symptoms. IV hydrocortisone 100 mg or dexamethasone 5 mg should follow. Severe cases should be admitted in case of relapse. Labile cases can be readily controlled by IV saline drip containing 1/500 000 adrenaline titrated against the patient's response. Milder cases can be discharged on a 5-day course of anti-histamines.

Adrenaline should never be given subcutaneously or intramuscularly as it is not absorbed in the collapsed patient, but will be when he has recovered. At this stage a minor overdose given in a moment of panic could be disastrous, especially in a hypertensive subject.

The need for subsequent desensitization must always be remembered and met. Cf. **Stings, wasp and bee**.

ANEURYSM, AORTIC

1 **Arch** — ascending, traditionally (but not reliably) syphilitic; descending, athero-sclerotic; may rupture or dissect. The latter is associated with agonizing dorsal pain and interference with peripheral pulses according to the anatomical situation. Rupture kills in a few moments.

2 **Abdominal**   (a) *upper* — severe lumbar pain and shock; (b) *lower* — poorly localized but severe pain in the pelvis, with shock and often interference with the peripheral pulses; may be symmetrical. Early recognition is important now that surgical repair is widely available and skilfully performed by many general surgeons. This can present with testicular pain of great severity.

ANKLE

Ankle sprains originate ·in four ways: inversion, overextension, overflexion, and eversion.

Inversion injuries

Commonly the lateral collateral ligaments are strained, or the subastragaloid joint sprained, or the antero-inferior tibio-fibular ligaments over-stressed. These injuries require only bandage support and active and energetic mobilization. The patient should be told to walk without a limp. It is better not to apply Elastoplast to start with in case of swelling or the need for subsequent X-ray. If pain or disability persists after the fourth day, continue double elastic support, e.g. Tubigrip, for 10–21 days. Ultrasound is helpful if used from the beginning. It relieves pain, reduces swelling, and permits proper movement, which is itself curative.

It is worth localizing the ligamentous lesion by palpation and manipulation in all cases.

Some sprains = partial rupture of the evertor mechanism, i.e. peroneal sprain. These are commonly associated with extensive bruising above and below the ankle and resolve slowly. Use supporting bandage (see **Bandages, supporting**).

X-ray is not necessary as a routine, but only where there is clinical indication or prospect of litigation against any third party. It is however increasingly difficult to refuse unnecessary requests for radiography in these litigious days. In the more rational communities of the Third World this is not so.

Overextension

can rupture both collateral ligaments and may be associated with compression fracture of the tarsal scaphoid. If it is combined with an axial force (e.g. in jumping from a height) talar fractures can occur and may well need internal fixation.

Overflexion

can be associated with 'third malleolus' fracture. Look for this on X-ray. Otherwise anterior capsular injury predominates and may be associated with unimportant flake-fractures from the neck of the talus.

Ruptured ligaments of the ankle (see Fig. 3)

Rupture of ankle ligaments leads to instability of the joint. It

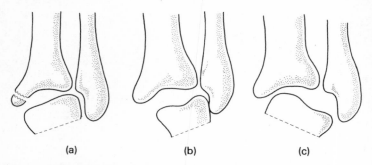

Fig. 3. Unstable ankle injuries demonstrated by forced inversion/
eversion under general anaesthetic.
  (a) Medial malleolus
  (b) Medial ligament
  (c) Lateral ligament

is clinically indicated by severe pain and swelling and failure to
respond to simple conservative treatment.

Rupture of the medial ligament needs surgical repair and is
demonstrated by stressed eversion radiography (under general
anaesthetic) (Fig. 3(b)). Opinions about treatment differ widely.

Rupture of the lateral ligament may be treated by POP for
6 weeks, or by surgical repair. It is demonstrated by stressed
inversion radiography (under general anaesthetic) (Fig. 3(c)).

Rupture of the inferior tibio-fibular ligament (tibio-fibular
diastasis) is readily shown by either stressed view, but is relatively rare. It generally requires internal fixation.

If there is a medial malleolar fracture it can be stable if the
periosteum (and thus the medial collateral ligament) is intact,
or it can be unstable as shown in Fig. 3(a).

Partial rupture of the lateral ligamentous/capsular complex
shows less marked talar tilt and heals more rapidly than complete
rupture. A shorter period (3 weeks) in POP is needed.

It is high time the best treatment for these conditions was
sorted out by adequate controlled trials. They are common
enough yet orthopaedic opinion is as variable as April weather.

Sometimes the interosseous ligament may be split as well. The fibular fracture may then be proximal (Maisonneuve injury).

ANTHRAX

Rare in the United Kingdom. Still a common zoonosis in the African continent and the Indian sub-continent. Occurs in hide and bone-meal handlers. There are three types:

1 **Cutaneous anthrax** (malignant pustule) Can be solitary and painless, or multiple and toxic.

2 **Pulmonary anthrax** — wool-sorters' disease. Highly toxic, haemorrhagic pneumonia.

3 **Intestinal anthrax** from eating infected meat. Fatal gastro-enteritis.

Diagnosis by bacteriological smears and guinea-pig inoculation.

Treatment by local measures dictated by general principles and IV infusion of penicillin-G 2 g daily till controlled. Streptomycin is alternative. Resuscitation for diffuse forms generally needed.

ANTIBIOTICS

These should be used as a matter of routine in the following cases:
Major injuries with soft tissue damage
Compound fractures
Major head injuries
Local sepsis which is not subsiding
Penetrating wounds of hand or foot
Together with surgery when abscesses are opened
All tendon injuries
Crushing injuries of extremities
Prophylaxis of infection in potentially infected wounds, e.g. hand wounds in meat and poultry handlers
In all drug addicts requiring anaesthesia for surgical procedures
They should not be used in the following cases:
Undiagnosed medical infections in children
Upper respiratory infections of viral origin
Minor superficial sepsis (impetigo, if severe, can be an exception)
Uncomplicated skin-wounds

It is sometimes difficult to decide which type of antibiotic to use, and the list below may give useful guidance. It is important to remember that a majority of staphylococcal infections is penicillin-

resistant (84 per cent in Hereford in 1983) and that a considerable proportion of streptococcal infections is tetracycline-resistant. Therefore:

1 In prophylaxis of infection in wounds and injuries give IM penicillin 600 mg and flucloxacillin 500 mg. If the wound is soiled, or prolonged prophylaxis is desirable, continue with oral Magnapen (= ampicillin + flucloxacillin) 500 mg four times a day before food.

2 In obviously staphylococcal infections (localized abscesses, whitlows, etc.) give cloxacillin or flucloxacillin 250 mg four times a day before food. An increasing number of hand infections show anaerobes on culture. Clindamycin may be considered the antibiotic of first choice especially if the pus is evil-smelling.

3 In all cases of cellulitis (always streptococcal until proved otherwise) give IM penicillin 600 mg, followed by oral flucloxacillin 250 mg four times a day before food. Oral penicillin is poorly absorbed in adults. Absorption of erythromycin is variable and unpredictable at all ages.

4 In cases of perianal or pararectal abscesses use clindamycin. See Abscesses.

5 In cases of penicillin sensitivity which would otherwise call for this class of antibiotic give cephradine 500 mg IM, followed by 500 mg four times a day after food by mouth for 5 days. Clindamycin (Dalacin-C) is sometimes kept in reserve. Some recommend it for all penicillin-sensitive patients (cf. **Abscesses**).

6 In cases of head injury where prophylaxis is called for because of suspicion or certainty of meningeal discontinuity give penicillin 600 mg, flucloxacillin 500 mg, and sulpha-dimidine 1 g, all IM.

7 It is worth recording the high effectiveness of metronidazole in anaerobic infections, particularly those due to *Bacteroides fragilis* (abscesses) and the combination of Vincent's spirochaetes and fusiforms which are the causative organisms of acute ulcerative gingivo-stomatitis. Recent advances in the management of abscesses and hand infections highlight the frequency of anaerobic infections. In general clinda-mycin is the best drug for these and more likely to be taken

Oral Magnapen has been replaced by Augmentin (= amoxycillin 250 mg + clavulanic acid 125 mg), which is cheaper, effective against resistant staphylococci, and has a better spectrum of effectiveness. Dosage suggested for prophylaxis: Augmentin 375 mg 8-hourly for 3 days. Treatment of severe infections: Augmentin 750 mg 8-hourly for 5 days.

Some of the newer penicillins may become established in this field, but their position and cost-efficiency is not yet secure.

Some of the third generation cephalosporins (e.g. Cefitoxime) given IV or IM may be better for single-dose treatments.

1983 saw the introduction of numerous new antibiotics which have hardly had time or outstanding features enough to make clinical recommendations clear.

than a complicated Augmentin/metronidazole regimen. Cf. **Abscesses**.

8 Clearly, subsequent bacteriological findings may modify these guidelines, but you must start somewhere.

The use of a single dose of antibiotic at the time of injury is intended to prevent the multiplication of externally entering contaminants. Practice suggests that it is effective provided the spectrum of cover is wide and the dose adequate. It is not likely to encourage the development of resistant strains as longer courses certainly will.

In using antibiotics the worst single crime you can commit is to use them ineffectively, i.e. in inadequate dosage or of irrelevant spectrum.

Penicillin sensitivity

Minor reactions are unimportant. Unpleasant sensations are controlled by antihistamines. They are not a contra-indication to the use of penicillin. These sensitivities are frequently greatly exaggerated by patients who are thereby deprived of the benefits of this unique antibiotic. If there is any doubt in your mind give an intradermal test dose of 0.1 ml of penicillin-G (= 5000 units benzylpenicillin) and observe the skin reaction after 15 minutes.

ANTISEPTICS

Tincture of iodine remains the only effective ånd persistent skin antiseptic which can penetrate the epidermis. It is a very important remedy in A & E work. Various preparations of povidone-iodine (USNF) are admirable antiseptics for wounds, but not for skin.

ARTEFACT INJURIES

must not be forgotten in diagnosis. Differential diagnosis is established by making the lesion inaccessible by appropriate bandaging or POP. Unravelling the underlying psychogenic noxa is no part of the casualty officer's job.

If a presenting condition doesn't make sense (e.g. apyrexial, red, hot swelling with normal white-count), suspect artefact.

This point of view has become increasingly acceptable in traditional surgical circles in the last two years. Recently published work extends the application to abscesses treated by the Leeds method. Cf. **Abscesses**.

Tincture of iodine is the only effective, superficial, emergency virucide and should be used generally for epidermal injuries, especially if caused by cats, rabbits, rats or other rodents, mad dogs, or humans. Rabid dog-bites should be treated with copious application of tincture of iodine as soon as possible. Cf. **Bites**, **Rabies**, Chlorhexidine is a useful antiseptic but does not penetrate skin and is not persistent.

ASPHYXIA

The numerous causes include:
    Laryngeal occlusion by food (see **Foreign bodies**)
    Laryngeal occlusion by swelling, secretions, or vomiting in
        the unconscious (see **Head injuries**)
    Severe acute asthma (q.v.)
    Drowning (q.v.) (water, blood, or vomit can drown)
    Facial injuries (see **Fractures**, Facial injuries)
    Acute pulmonary oedema (see **Asthma, cardiac**)
    Chest injuries (q.v.)
    Anaphylaxis (q.v.) (e.g. wasp-sting sensitivity) (See **Stings, wasp
        and bee**)
    Angio-neurotic oedema (cf. **Anaphylaxis**)
    Epiglottitis (see **Croup**)
    Laryngeal oedema (croup (q.v.) etc.)

ASSAULT, PHYSICAL

Serious assaults often lead to multiple injuries and the victim is
often drunk, or a bad witness, or both. If in doubt, admit. Injuries
from assault are a growing epidemic in Britain.

ASSAULT, SEXUAL

You may be asked by patient, relatives, or police to examine such
a case. The patient must be stripped completely and notes taken
carefully of every injury and mark on the skin. Nails are
particularly important: scrapings from under a girl's nails may
give vital evidence. Note general appearance, clothing, stains,
soiling of garments, emotional state. Take the following specimens
to examine for spermatozoa:
    Moist swab from perineum
    Moist swab from vulva
    Moist swab from vagina
    Moist swab from pubic hair
    Clippings of pubic hair, into a sterile bottle
If possible give all specimens to the police for examination at a
forensic laboratory.
    Always have a chaperone/witness.

ASTHMA, BRONCHIAL

Essentially a complaint for GP management and supervision; but from time to time severe attacks of breathlessness reach accident departments. They are dealt with by

1 **Salbutamol nebulization** 0.5 ml of respirator solution in 2 ml of normal saline (5 mg/ml) for children under 5 years; 1 ml of respirator solution in 5 ml of saline for children of 5–10 years; 2 ml of respirator solution in 5–10 ml of saline for adults. Salbutamol is a β2-adrenoceptor stimulant.

This is outstandingly the best emergency treatment. The solution is put into the receptacle of the nebulizer attached to a standard oxygen mask and nebulized by the hospital oxygen supply. Its action is gentle, rapid, and efficient. The rate and quantity of the nebulization is determined by its effects. Nebulization can also be done by electrically driven air-pump, or by air compressed by foot-pump. Various models are available. Oxygen is best.

2 **Salbutamol by IVI** 50 gm/ml (up to 250 gm) given slowly and titrated by its effects. This can also be given by saline infusion (1 mg/ml solution) – 5 ml in 600 ml of normal saline.

3 **Aminophylline** (a xanthine bronchodilator) by IVI, 250–500 mg given slowly over 10–15 minutes. This has been largely superseded by the above, but may still be available when salbutamol is not.

4 **Steroids** by IVI have an important adjuvant role in allergic or anaphylactic asthma. They are secondary to the above remedies as they are slow-acting. Hydrocortisone sodium succinate 200 mg IV is appropriate. Onset time 2–4 hours.

5 **Adrenaline** (1 ml of 0.1% adrenaline tartrate or 10 ml of 0.01% adrenaline tartrate) is a time-honoured and widely used remedy, current at any rate since Sir Arthur Hirst's heyday in the 1930s. It is now in eclipse. If it is to be used it must be given IV 0.1 ml at a time and never subcutaneously where it can linger as a deposit to be released quite unpredictably in time and quantity as the failing circulation recovers. I never use it. It is popular in the USA.

6 **Adrenaline by IVI** or as an infusion in normal saline titrated against the patient's response (rising pulse-rate and muscular

tremor are the two indices of maximal dosage), 1 mg as a bolus or 10 mg in 600 ml normal saline infusion.

This remains the treatment of choice for acute anaphylaxis (q.v.) with or without asthma. See **Anaphylaxis, acute**.

7 **Sedation** is rarely called for but may have to be resorted to in the violently anoxic patient who fights off the mask and needle alike. IMI Midazolam 5—10 mg may bring a desperate situation under control. It is a mild respiratory-centre depressant.

ASTHMA, CARDIAC

An acute pulmonary oedema due to left-sided heart failure, usually relieved by sedation with Cyclimorph 15 mg IV (specific vaso-dilator as well as sedative), and IV diuretic (e.g. frusemide 50 mg). Digitalization may be needed too, especially if acute atrial fibrillation is the cause: IV digoxin 0.5 mg × 2. Repeat frusemide to a maximum of 200 mg in 50 mg aliquots. Some recommend amiloride hydrochloride as a more effective diuretic because of its vasomotor effects.

BACKS

Half the population suffers from backache, owing to misuse or disuse (the latter more prevalent). The other half is still liable to back injury or disease. The commonest manifestations are:

1 **Acute lumbago** See **Lumbago**.

2 **Chronic back-ache** may well need investigation of lumbar spine and sacro-iliac joints, gynaecological apparatus, and psychosexual background. Treatment: once organic disease is excluded, vigorous mobilization and extension exercises. A day's digging or some all-in wrestling is excellent. Manipulation is good in experienced hands. Not strictly an emergency commitment however.

3 **Prolapsed intervertebral disc** should not be diagnosed without symptoms or signs of root-pressure, e.g. limitation of straight-leg raising or other root pain. The severe, acute prolapsed intervertebral disc is usually well dealt with by means of POP jacket — 3 weeks in the first instance. The patient should be warned that it may be 10 days before the

Note: Terbutaline (a selective $\beta 2$-adrenoceptor stimulant) is similar to salbutamol and can be given by IMI. It is highly regarded in the USA.

root pain subsides. If the pain is intolerable, admission for rest and traction is occasionally required.

4 **Ligamentous injury**, especially interspinous ligament injury: local tenderness acute. Local Marcain (bupivacaine) 0.5%, 5 ml at least, and mobilization.

5 **Quadratus lumborum myalgia** (with or without exertion fracture of the twelfth rib) Infiltrate with local marcain 0.5% 10 ml, and mobilize. Avoid renal biopsy.

6 **Dorsago** An acute dorsal syndrome involving the erector spinae or serratus posterior, comparable in its characteristics to acute lumbago; is common and can be very painful and disabling. Infiltration of local around the relevant costo-vertebral or interfacetal joint can be dramatically helpful. Painful muscle spasms are particularly distressing.

Back injuries are dealt with under **Fractures**.

Pathological fractures, Pott's disease, and other medical mysteries should always be borne in mind but are not strictly an A & E commitment.

Physiotherapists have much to offer to acute cases of lumbago and to the re-education of the backs of recurrent sufferers. See **Physiotherapy**.

BALANITIS

Can produce acute urinary retention. This normally responds to frequent hot baths, and antibiotic eye-ointment applied inside the prepuce. Antiseptic ointment would be better but it is not available in any appropriate form. In urgent cases dorsal slit may be required — easily performed under lignocaine 1%. The subsequent need for formal circumcision should not be forgotten, and careful follow-up is important. Fungal balanitis also occurs but is seldom acute.

BANDAGES, ICHTHOPASTE AND VISCOPASTE

These are a very useful and comfortable application for wounds and skin lesions in atrophic skins, or in the presence of poor blood-supply in the leg. They should always be used two at a

If precise localization is impossible inject 2–4 boluses of 5 ml bupivacaine 0.25% deep into the paravertebral muscles. This will generally control the painful spasms.

time and always from the toes to the knee. They can often be left
in place for 1—2 weeks at a time. After application the patient
should be asked to rest the leg for an hour. The bandage then
'sets' without wrinkles. Ichthopaste can often be used where
there is infection or an eczematous skin, in combination with topical antiseptics and saline bathing, and is generally the more useful
of the two for this kind of patient. It is softer (*Cont. opposite*)

BANDAGES, SUPPORTING

It is frequently desirable to use supporting bandages for the
following purposes:
  1 Control of oedema
  2 Support of injured bones and joints
  3 Improvement of blood-supply in the lower limb
The available types of support are
    Tubigrip
    Crepe bandage
    Elastoplast
    Zinc paste
Zinc paste is messy but effective for 1 and 3, but not much use
for 2. Crepe bandage is useful for only 4 hours and then needs
expert reapplication. Elastoplast is good for all three, but tiresome
to remove, and often produces skin reactions. It is also partially
radiopaque. Tubigrip is as good as any for all three, and has no
disadvantages. All patients should be told to remove it at night if
it is uncomfortable, and reapply it before rising. It is no more
expensive than Elastoplast. Don't forget that *all wounds of the
leg and foot require toe-to-knee support.*

BATTERED BABIES AND OTHERS

Children and spouses have been battered by children, parents, and
spouses since Adam and Eve got their notice to quit. The fashionable hue and cry after these cases makes it important for all
accident staff to be alert in watching for signs of such cruelty. To
this end a register of injuries to children under five may be kept.
The hospital social worker may keep an eye on it and follow up
cases of recurrent injury. The simple rule to follow is to admit

(*Cont.*) and more comfortable than plain zinc-paste bandages, besides containing ichthammol as a stimulant to epidermal growth; it is a very useful dressing for slow healing hand and finger wounds, especially where the volar skin is hard and thick.

Trade names used here are of long established makes which have led the field for many years. Many other brands are available now, some good and some not so good.

every suspected case of 'battering' and inform the senior casualty officer on duty and the paediatric consultant personally of your suspicions, as well as writing clearly on the casualty card 'Suspected case of non-accidental injury (NAI): multiple/recurrent injuries of unsatisfactorily explained origin'.

Parents may give plausible explanations of each incident. The child generally shows signs of emotional deprivation; in cruder cases fear and withdrawal are obvious. It is no part of a casualty officer's duty to unwind this tangled skein, but *admission* is essential and the paediatrician must be personally notified ·of your suspicions in the matter. There are all too many cases on record where tragic and avoidable death has befallen the poor child simply through lack of communication.

BITES

Dog

Bruising often more significant than laceration. Tetanus cover essential. Penicillin and cloxacillin IM if in doubt. Tincture of iodine remains the only effective and long-acting skin antiseptic. Sensitivity to it is very rare and easily counteracted by topical steroids when it occurs. The common dog-bites of the face in children need early and careful repair (often under general anaesthetic) with subsequent scar-revision if needed.

Rabies (q.v.)

Phone virological laboratory. Phone police and community physician if there is a definite suspicion.

Cat

Needs to be treated with extra respect because of the risk of cat-bite/scratch fever (cf. **Viruses**). Tincture of iodine (the only effective percutaneous protein-coagulant which may kill an invading virus) and penicillin and cloxacillin IM.

Monkey

Can be nasty. Treat as cat.

In the present social climate battered spouses and parents are seen with increasing frequency, and can be equally distressing to deal with and difficult to help. Every case needs time, care, and gentleness.

Pasteurella septica/multocida: an important pathogen from dog and cat bites. Sensitive to penicillins and methycillins.

Pig

Pigs have very strong jaws and can inflict a lot of tissue damage and/or fractures through undamaged skin and clothing. Treat all symptoms with respect.

Horse

Superficial nip usually. A recent case where a jealous stallion grabbed his handler by the scrotum, picked him up and shook him, then dropped him on his back, was exciting but unusual.

Snake

The adder is the one poisonous snake native to Britain, and the only one which bites. Bites generally produce local symptoms only, and disproportionate fear. Snake-bite serum (prepared from a different species anyhow) should not be used automatically. Treatment is symptomatic and, in the rare cases of general poisoning with collapse (histamine-type symptoms), admission, IV fluids, and general supportive treatment are necessary. See **Poisoning**. To patients genuinely collapsed at first contact give IV adrenaline 0.1% 0.1 ml per minute, and IV hydrocortisone 100 mg or dexamethasone 4 mg.

There is no obtainable consensus about the use of antivenom in Britain. Local infiltration of this antidote is recommended by Reid (*BMJ* ii 153–6 (1976)) to prevent tissue necrosis. A symposial recommendation (*Prescriber's Journal* 19 No. 6 190–9 (December 1979)) seems sensible, namely to give IV antiserum in severe systemic poisoning, as indicated by coma and spontaneous bleeding. High peripheral leucocytosis, abnormal ECG and raised serum creatine phosphokinase support the decision. Initial dosage of Zagreb antiserum is 2 × 5.4 ml in 100 ml of saline given during 1 hour. Reactions are said to be rare with the highly purified Zagreb antiserum which is kept by most district hospitals, but should still be looked for and treated if they arise.

In other countries (and in zoological collections) antiserum is of prime importance but often less available if only because of distance.

Indian cobras and kraits, African mambas, coral snakes, and spitting cobras, Australian sea-snakes and other elapids, and

Actinobacillus lignieresii from horses and cattle can infect man. Consult bacteriologist.

An interesting personal account of the experience of bite by a sand viper (*V. ammodytes*) by D. F. Jackson gives a useful insight into the symptoms of snake-bite and its relative importance as an infrequent and often minor injury (*Lancet* ii 686 (1980)).

TABLE 1 SNAKE BITES

Snake genera	Snake species	Features	Distribution	Incidence of envenoming after bite	Local effects	Systemic effects	Mortality	Survival time
Elapidae	Cobras Mambas Kraits	Short fixed fangs	World-wide except Europe	50%	(Cobras only) slow swelling and necrosis	Neurotoxic: ptosis, dysphagia, respiratory palsy, cardiac palsy	5%	5–20 hours
Hydrophiidae	Sea-snakes	Flat tails, short fangs	Asian–Pacific coastal waters	20%	None	Mytoxic: myalgia on movement, paresis, myoglobinuria, hyperkalaemia	10%	15 hours
Viperidae	True vipers Carpet viper Russel's viper European adder	Long hinged fangs	World-wide except America and Asia	30%	Rapid swelling and necrosis	Vasculotoxic: abnormal bleeding, loss of blood clotting, shock	1%	2 days
	Crotalinae (pit vipers) Rattlesnakes Fer de lance Malayan viper	Small heat-sensitive pits on head	Asia and America					

.

various different vipers from many different countries call urgently for antiserum. Whether this is available readily or not, it must never be forgotten that sound supportive treatment with large doses of hydrocortisone will often save lives.

The supply of multivalent anti-venene is general in the Indian sub-continent, but only in certain areas of the continent of Africa. Cf. **Collapse and coma.**

Table 1 (leaning heavily on H. A. Reid of Liverpool), p. 23, summarizes the basic data.

Man

Often met with after fights; aimed at inflicting maximum damage to ears, noses, fingers, etc. Treat on general lines.

Insect

Even the angriest looking swelling is generally an alkaloid reaction or an allergic one. Secondary infection is rare and generally accompanied by lymphangitis. Therefore an antihistamine is the first treatment to try. See **Stings, wasp and bee.**

BLISTERS

Whether due to burning, friction, or unknown causes, blisters provide a good sterile dressing, and are best left for a few days. If they burst or merely persist after 5 days they should never be pricked, but removed in the entirety. Appropriate treatment can then be given to the deeper tissues disclosed by their removal. Pricking can only introduce infection into a sterile medium ideally suited to the multiplication of bacteria.

Cold epidermolysis often puzzles people. It presents as a brownish blister of the sole in cold or damp winter weather. Sometimes this becomes painful. If recurrent, early injection of intradermal Depo-Medrone will abort it.

BURNS

First aid

The overriding aim of treating **first degree burns** is to stop them

Spider

A useful review article (*Lancet* **ii** 133 (1980)) on spider-bites largely exonerates *Lycosa tarantula* from seriously morbid effects and directs attention to *Latrodectus tredecimguttatus*. The female of the latter species is black or brown with usually 13 (as its name implies) red spots. This is the black widow and its bite is said to give 5% mortality. Muscular cramps and hypertension are characteristic and a specific antivenom is available. Otherwise treatment is supportive.

becoming second degree. Do this by spraying with Terracortril. This produces rapid blanching and reduction of the quantity of histamine produced locally. Some disapprove of this use but 17 years' experience has shown no disadvantages.

The overriding aim of treating **second degree burns** is to stop them becoming third degree. Do this by spraying with Terracortril. Follow by using home treatment with antiseptic cream and no dressing or other covering to the burns. Fluid replacement should be started early in a patient with extensive burns (especially second degree). Plasma (PPF) is the specific as it is serum that is lost, but saline is better than nothing and Haemaccel nearly as good as plasma.

If less than 20 per cent body area is involved treatment as out-patient using the Flamazine/exposure technique is successful (see 'Subsequent treatment' below). Healing time is reduced, scarring and disability minimized, and hospital attendances decimated. In children under 12 years 10 per cent body area is regarded as the critical limit above which admission is generally required because of the risk to their fluid and electrolyte balance.

Third degree burns (full thickness skin death, characterized by loss of appreciation of pinprick) are susceptible to similar treatment and consideration. Usually fluid/electrolyte loss is less than in second degree burns. But other considerations of healing and skin-replacement arise. A rational policy of management should be worked out in collaboration with any available or interested (plastic) surgeon.

In general out-patient treatment is always to be chosen if possible. Hospital treatment of burns is bedevilled by the ineluctable development of widely resistant bacterial strains, which can readily prove damaging or fatal in otherwise curable cases.

Subsequent treatment

Burns must always be treated with exposure if possible. The currently (1984) best (out-patient) application is the silver sulphadiazine cream (Flamazine) developed at the burns unit of the Birmingham Accident Hospital. This should be applied at least twice daily with a clean finger. No dressing at all should be used, and the patient should be nursed night and day in a

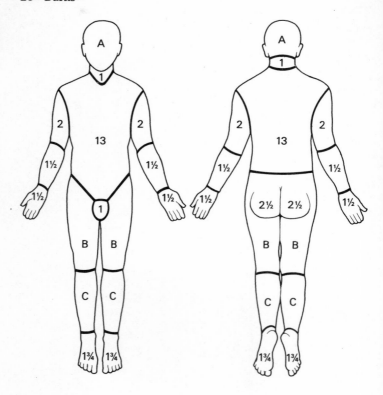

Relative percentage of areas affected by growth

Age in years	0	1	5	10	15	Adult
A – ½ of head	9½	8½	6½	5½	4½	3½
B – ½ of one thigh	2¾	3¼	4	4¼	4½	4¾
C – ½ of one leg	2½	2½	2¾	3	3¼	3½

Fig. 4. Estimation of percentage of total body area burns in children by Lund and Browder charts.

warm room without covering, and without dressing; for small burns dibromopropamidine cream (Brulidine) is sufficient. If dressings are inescapable, as in children's nappy area, chlorhexidine-impregnated tulle (Bactigras) is a rational and effective application. This is a developing area of treatment and new approaches should be critically assessed. At all events detailed written instructions and regular after-care should be insisted upon in all cases.

Specialized burns units have to have different approaches to these problems but are few and far between. They are generally able to accept only severe cases.

After-care

Burns characteristically heal with a thin epithelial layer which dries and cracks, admitting infection and delaying full healing. *Grease* is the answer and should always be prescribed. Hibitane cream is a useful application. Post-healing pruritus can be a problem with severe burns. Topical steroids help a bit but must be used for a limited period only.

Hydrofluoric acid burns (steel industry mostly)

First aid   Calcium gluconate gel 1.5% by inunction locally

Second aid   Excision and irrigation with cetrimide solution if of full thickness

Eyes   Irrigation with warm sterile water followed by instillation of sterile calcium gluconate solution 10% w/v

Bitumen burns

are often severe but the bitumen supplies a sterile non-antigenic dressing. It is better not to attempt to remove it except by slow solution in liquid paraffin. Forcible removal only leads to further injury.

Degrees of burning

A useful classification, currently unfashionable. Defined as follows:

First degree   Erythema without blistering

Second degree   Blistering (i.e. epidermal loss) without full thickness skin damage

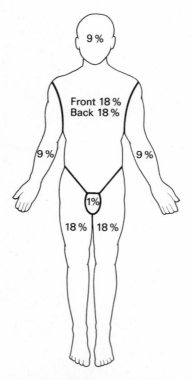

Fig. 5. Simple 'rule of nine' for rapid estimation of total body area burnt in adults.

Third degree Full thickness skin damage but not necessarily destruction of the whole of the dermis

Many people prefer to classify burns as either 'superficial' or 'full thickness'.

BURSITIS

The treatment for this condition is *rest* and protection. In acute olecranon or pre-patellar bursitis, even when grossly inflamed, the fluid contents are generally sterile. *Primary aspiration is in all*

severe cases absolutely contra-indicated. It all too often results in either infection or a chronic sinus or both. In the subacute or chronic phase aspiration and introduction of Depo-Medrone is helpful, and may materially shorten the disability and pain. Puncture should be made at the least dependent point. Full sterile precautions in theatre should be observed. A purulent bursitis is a rarity and generally the outcome of meddlesome treatment. If it occurs it is still best treated conservatively with antibiotics and rest.

Pre-patellar bursitis settles rapidly in a well-padded plaster cylinder.

Recurrent bursitis should be referred, failing all else, to the GP for formal surgical removal of the bursa to be arranged, or directly to the surgical specialist if you are permitted to do this. Don't excise bursae yourself as healing over bony prominences (e.g. the olecranon) is often very unsatisfactory. Leave this to a specialist.

CARDIAC ARREST

External cardiac massage and positive pressure ventilation – 4 compressions to each ventilation – until the d.c. defibrillator is lined up; then give a 200 J shock increasing by 100 J to 400 J. Ask someone else to apply the electrode jelly and get everyone to stand clear. It is best if no one but the patient gets the 400 J. Give concurrent intubation and oxygen from the anaesthetic machine, and call the medical registrar to take over continued supervision. In the absence of regular cardiac action potentials on the monitor, if you have one, fixed dilated pupils unresponsive to light are a good indication that it is time to give up. Cardiac arrest is generally effectively reversed if it occurs in the accident department or intensive care unit; if it occurs elsewhere, very rarely. Ward treatment of arrest can be successful if monitor and defibrillator are ready and at hand in every ward always, and all the nursing staff are familiar with their use and know when to use them. Cf. **Myocardial infarction.**

Biochemistry

There is no general agreement about the biochemical support needed in the acute phase of cardiac arrest and for myocardial

Cardiac pacing by oesophageal catheter (Vygon — 'oeso-cath') offers a hopeful advance in the management of patients with severe burns or multiple injuries complicated by arrhythmias or arrest. There is much work to be done in adapting this method for monitoring such patients, and for using it in asystole preceded by intracardiac atropine 0.6 mg × 2. For asystole see also p. 109a.

infarction (q.v.). It has been suggested that every case of arrest should have 100 mEq. of bicarbonate IV and that if adrenaline is used it should be given with calcium gluconate 10% 10 ml. So far as the primary treatment of arrest is concerned, electricity is the first need. Drugs may follow it but should never precede it. When the dust settles give a 50 ml bolus of 8.4% bicarbonate solution.

In my own experience and opinion it is best to discourage junior staff from using β-blockers and adrenerges which carry considerable dangers if used injudiciously. They should however be kept available for visiting physicians who can understand their bewildering side-effects.

Expert support

Most hospitals have an emergency call system for arrests. It is a good and useful service but no substitute for prompt and effective action by the person on the spot.

External cardiac stimulation

Also called cardiac massage or cardiac compression. Eminent counsellors give quite astounding advice on this subject. If you squeeze a heart in asystole you can produce a shock wave which is similar and synchronous in the venous and arterial systems, but it is doubtful whether you can produce circulation or oxygenation of the blood.

If you try too vigorously to compress a heart externally, sufficiently to empty it (if this is possible), you will either fail or break numerous ribs and possibly the sternum. You are only administering a brutal and useless trauma to a damaged organ 'in articulo mortis'. Autopsy reports after external cardiac stimulation retail instances of inexcusable injury to the heart, lungs, and other intra-thoracic contents. Even the liver has been ruptured.

If you give a series of short, sharp (but moderate) taps to a fibrillating myocardium you may induce it to revert to coordinated electrical activity. You may also produce single uncoordinated contractions of the ventricular muscle in asystole. You may save a life. It is better to do this than to terminate one by a violent and useless assault.

Recent work on cardiogenic shock draws attention to the use of dopamine to combat hypotension. Its role is not yet precisely defined but do not forget its use; e.g. Timmis and others, *BMJ* **282** 7 (1981).

This treatment, properly given, *must* be accompanied by alternate positive-pressure ventilation — 4 compressions to each ventilation.

The compressions should be delivered at 1-second intervals by a bouncing thrust with the heel of the hand over the sterno-xiphisternal joint. You should practise on the living, who will tell you if you are hurting them. If you are, you are being too rough. The force used should be sufficient to 'knock the wind out of them'.

The bounce must be applied with a force appropriate to the age of the victim and finger-tip pressure is enough for little babies.

You may never need to do this if you have an electrical defibrillator, but please teach others concerned — especially ambulance crews and energetic first-aiders — the truth of the matter.

Whether this account of the process is veridical or not is in dispute and likely to remain so. Objective evidence is barely obtainable in the circumstances of urgency surrounding cardiac arrest and conclusive observations completely unobtainable.

CARDIAC FAILURE, ACUTE CONGESTIVE

See **Asthma, cardiac.**

CARDIAC TAMPONADE

Difficult to diagnose, often missed, and therefore fatal. If it is spotted it may be curable. It can be due to crush injury leading to bursting of the myocardium, or penetrating wounds from external (metallic fragments) or internal (rib penetration) causes. Not an uncommon steering-wheel injury in car accidents.

Clinical diagnosis   Accelerating pulse and falling blood pressure in the absence of evident blood-loss. Sudden cardiac arrest ensues and is usually irreversible. Rapid rise of JVP and JVF.

Confirmatory diagnosis   Pericardial paracentesis using a wide-bore needle into the 4th intercostal space at the left sternal margin.

Treatment   Urgent surgical repair.

PPV: the current fashion (1984) is for concurrent PPV and ECM without alternation. It works well when tempered with common sense.

If in doubt try squeezing blood out of the heart of a newly killed pig.

'Medical' haemato-pericardium due to ruptured scar of previous infarction, or to dissecting aneurysm of the ascending aorta, is for practical purposes a terminal event.

If it is one of many injuries cardiac tamponade is certain to be missed unless you do a routine paracentesis of the pericardium in all such cases.

CARDIO-VASCULAR EMERGENCIES, OTHER

Cardio-vascular collapse can also be caused (especially in elderly people with poor cardiac reserves) by dissecting *aneurysm* of the aorta or major peripheral *embolism*. Now that vascular surgery is widely available early diagnosis and referral are urgent. The collapsed patient with absent femoral pulses may have a saddle embolus at the bifurcation or a low dissection of the aortic wall. Femoral embolism, or even popliteal, can present with collapse as well as signs of arterial occulsion, and both may be amenable to urgent surgery.

Heart-block is readily diagnosed clinically and by ECG but intermittent heart-block with occasional Stokes—Adams syncopal attacks may be undetectable except by 24-hour ECG.

Rare causes include visceral embolism (especially mesenteric), intermittent mitral obstruction (due to intra-atrial thrombosis or pedunculated myxoma), acute toxic myocarditis (e.g. after influenza), and, among those suffering from starvation, beri-beri heart.

Pulmonary embolism should always be borne in mind even in the absence of evidence of previous DVT, but it rarely presents in A & E departments in my own experience.

Cf. **Myocardial infarction** and **Collapse and coma.**

CAUSTICS

Strong acids make painful burns on the skin and so generally get rapid treatment. Strong alkalis and phenols, unfortunately, are generally painless and may not be recognized as caustic. As a result their caustic onset is insidious and more disastrous. Its full extent may not be recognized for several weeks — e.g. on the back of the hand, where primary tendon damage may pass unrecognized. When children drink any of the many caustics in the home (e.g. alkaline domestic cleaners) there is a strong case for considering admission to a chest unit where oesophagoscopy can if necessary

be followed by operation to close mediastinal leaks from the oesophagus. See also **Poisoning**.

CHEST INJURIES

Penetrating and crushing injuries of the chest are always serious and can be rapidly fatal if not dealt with expeditiously. Perforations or lacerations of the lung can rapidly embarrass respiration by haemothorax, tension pneumothorax, or a combination of both. Early intercostal intubation is safe and harmless. It should be done always and only in the second intercostal space in the mid-clavicular line. The lateral approach should never be used in an emergency as there may be a ruptured diaphragm with liver or stomach in the chest. The transparent trochar-mounted tubes are best. They need not be very wide in the acute stage; 8 mm suffices.

Flail segment, noticed by observation of paradoxical respiration, may be extensive and rapidly produce respiratory embarrassment especially in aged or emphysematous casualties. Pressure on the free segment and endotracheal intubation solves the problem by enabling positive pressure ventilation.

Always feel for *surgical emphysema* and look for it on X-rays. It draws attention to injuries of lungs or bronchi which may otherwise be missed. Chest radiographs in the erect position tell much more than supine ones.

Mediastinal bleeding may follow from compression injuries of the chest, especially in the elderly in whom quite small arteries, brittle from degenerative changes, may bleed slowly and unnoticed. The signs are of rising pulse and falling blood-pressure, and rapidly increasing JVF and JVP, in a chest-injured patient who shows no apparent blood-loss and normal lung fields. There may be no observable fracture. There is, however, progressive widening of the mediastinal shadow on serial X-ray. The internal mammary artery is usually the culprit and it is simple to tie it off if the condition is recognized. If it is not the victim dies.

Shock-lung is a condition due to major blood-loss in which blood-volume replacement has been inadequate or delayed. It is comparable to renal shut-down; the basic lesion is disseminated intra-capillary coagulation (DIC). There is considerable evidence that

The causative role of surfactant disturbance is not yet clear.

large doses of steroids materially reduce the severity of the DIC and its consequences. Medrone (methylprednisolone sodium succinate 2 g) should be given at the earliest opportunity IV if there is any likelihood of its development.

Lung contusion is due to severe compressing forces of brief duration on a relatively elastic chest wall; there are not necessarily any associated fractures of the thoracic cage. It may initiate DIC and warrants the same treatment.

Mediastinal compression, due to more gradual compression of the upper mediastinum, is recognized by the development of petechial haemorrhages in the skin of the chest and shoulders above the line of compression (e.g. by tractor wheels, accidents in civil engineering diggings, or compression by buckled road-vehicle frames). Snowstorm petechiae in the lungs also develop and may produce symptoms.

The same treatment may be worth using and is apparently harmless.

Blast-lung (pulmonary barotrauma) is also associated with DIC. Steroids are well worth trying, together with all other supportive care available.

Serious chest injuries need early transfer to intensive therapy. If there is close collaboration between A & E and ITU it may be desirable to estimate blood-gases in A & E (pO_2 and pCO_2). If not they are best left till later.

Normals: Blood   pH (7.35–7.45)
pO_2 (9–15 kPa)
pCO_2 (4.5–6.1 kPa)
base excess (± 2.5 mmol/l)
standard bicarbonate (21–26 mmol/l)

CHILDREN

Dealing with injured and acutely ill children is part of the art of medicine and depends upon the quality of the person concerned and his attitude of mind. Experience and example can refine these basic attitudes, but there is one small point of technique which is worth recording. When you first see such a child pay no attention to his injury or illness but only to him as a person,

Methylprednisolone up to 2 g for **multiply and seriously injured** patients can be given with profit as early as possible in the first unit of IV fluid or blood. If given as an IV bolus it usually induces nausea and vomiting.

starting with the customary greeting and moving gently on via discussion of matters of interest, such as his clothes, toys, occupation and so on, to the purpose of his visit. This does not apply when there is severe pain which needs early relief before these courtesies can begin to serve their purpose of establishing an *ad hoc* relationship. Special aspects of the care of such casualties are admirably dealt with by Illingworth. See **Bibliography**, Miscellaneous, p. 170.

CLINICAL TRIALS

The benefits to an accident unit of regularly conducting limited prospective clinical trials are immense. The patients benefit from the instant rise in interest that is generated and the staff (medical and nursing alike) proportionately. The practical benefits that arise are not inconsiderable either, in an area where there are wide and fundamental disparities in everyday practice. For instance, a simple clinical trial was set up to test the hypothesis that there is no benefit to be gained from treating simple hand wounds needing suturing with prophylactic antibiotics. The results obtained in a group of patients who were so treated were identical to those obtained in a control group who were not. The conclusion that prophylactic antibiotics do not have any significant effect is an important one. Provided that the trial is limited in its scope, and carefully designed to test the minimum of variables (if possible two only), clear and statistically significant results can be obtained which allow objective improvements in the rational care of injured patients. The accident field offers endless scope for this elementary and satisfying activity, which should be encouraged everywhere provided that patients are helped by it without being put to unjustifiable inconvenience, and adequate supervision and control of the design and management of the trials is available. (See letter from Beesley, Bowden, Hardy, and Reynolds. *Injury* 6 366 (1976).)

CLINICS

Make sure you know what clinics are available and when. There will be a notice-board somewhere in your accident unit with the necessary details. Most hospitals have special clinics for:

Longer-term research projects in A & E units are of even greater value in maintaining the intellectual and scientific acumen and enthusiasm of senior staff, as well as making a potential contribution to more effective patient care. They should be encouraged in every way possible. Computerization of records will produce invaluable opportunities in this field.

Fracture follow-up
Hand injuries
Lumps and bumps
VD

There is space opposite to record times and places.

Some hospitals encourage direct referrals to specialist medical, surgical, orthopaedic, dermatological, otorhinolaryngological, or what-not clinics. Some insist on referral via the patient's own doctor. Find out the policy on referrals, then act accordingly.

COLD INJURY

Cold epidermolysis

See Blisters.

Frost-bite

Peripheral gangrene usually of fingers or toes due to prolonged circulatory shut-down and interstitial ice-formation with irreversible tissue damage. A surgical problem. The slowest possible re-warming is essential.

Hypothermia (q.v)

Trench foot

Cold immersion in mud or water without freezing produces ischaemia; re-warming causes a hyperaemic red and painful extremity responding to the accumulated tissue metabolites of the ischaemic phase. If this situation is recurrent or prolonged tissue death arises.

A general recommendation is to warm ischaemic tissue dry and slowly, to cool painful hyperaemic tissue, preferably with a fan, and to avoid scrupulously any trauma to tissue-damaged areas by pressure or friction.

After the primary treatment (which is largely aimed at preventing further damage) general surgical principles prevail.

COLLAPSE AND COMA

Collapse is an overworked label which includes all sorts of emo-

It is important to have routine follow-up clinics run by a senior member of staff, preferably a consultant, in order to provide a supervisory function and to keep the number of return visits to a minimum.

tional upsets from terror to tantrums. Coma is a more definite condition of inaccessible loss of consciousness, but the two overlap and are more easily dealt with together. Physical causes include:

Cardio-vascular collapse

Due to internal or external bleeding, e.g.:

Oesophageal varices (Sengstaken's tube for control)

Peptic ulceration

Bleeding Meckel's diverticulum

Blood-abnormalities or dyscrasias

Telangiectasia of the colon and various other abnormalities

Spontaneous rupture of a varicose vein in the lonely aged can exsanguinate rapidly. There may be little external evidence of it on arrival at the A & E unit. The evidence is generally left on the bedroom floor.

Blood-volume replacement urgently with Haemaccel and then with cross-matched blood or concentrated red cells meet the emergency.

A ruptured varicose vein only needs elevation of the limb to staunch the bleeding and make surgical closure simpler. In general under-running of the leak with a Dexon suture or two is the first-aid measure of choice. Traumatic haemorrhages are dealt with elsewhere in more detail.

See **Accidents, major, Chest injuries, Gun-shot wounds, Shock, surgical**; cf. **Epistaxis, Haematological emergencies.**

Primary pump failure

Due to:

Myocardial infarction (q.v.) — far the commonest

Left ventricular failure especially with aortic stenosis

Dysrhythmia — especially heart-block leading to Stokes—Adams attacks, but also ventricular arrhythmia (beware of the β-blockers) and acute atrial fibrillation

Acute thyrotoxic crisis (cf. p. 53a)

Paroxysmal tachycardia

Acute ventricular failure in the pulmonary cripple who blows an emphysematous bulla and develops a pneumothorax

Atrial myxoma acting as intermittent ball-valve

Vertebro-basilar insufficiency (pulpit watchers' syndrome)

Posturally associated syncopal attacks due to a combination of athero-sclerosis and cervical spondylosis.

Respiratory crises

'Café coronary' Bolus impaction in the laryngeal introitus. Treated by the 'piston punch' in the epigastrium or the 'brisk bear-hug' applied in the same area. These manipulations (Heimlich's manoeuvre is easier to demonstrate than to describe) are intended to produce a sufficient expiratory explosion to dislodge the bolus. Cf. **Asphyxia.**

Subacute pneumonia in the aged and **breath-holding attacks** in infants.

Acute bronchiolitis in small children may present with collapse and loss of consciousness. Acute laryngo-tracheo-bronchitis may produce early impairment of consciousness.

First aid is always to establish an airway if possible and to give oxygen. Complete asphyxia may need laryngotomy (q.v.).

Neurological crises, especially intracranial bleeds, whether intra-cerebral or subarachnoid. See **Neurological emergencies.**

Epilepsy with its puzzling post-ictal confusion. Diazepam as first line treatment, IV or IM — 10 mg is a good initial dose.

Self-poisoning and accidental poisoning   See **Poisoning.**

Carbon monoxide poisoning is now mostly found in closed garages, cars stuck in snow-drifts with their engines running to keep the occupants warm, and closed rooms with leaky stove-pipes.

Town gas-supplies in Great Britain are now only harmful in so far as they can displace the oxygen content of a closed room and so produce asphyxia. In some other countries coal is still coked to produce carbon monoxide.

Various types of asphyxia   A vivid recollection is of a married couple who went to bed, in a small room with no ventilation and a paraffin heater, and were brought in one bitter winter's morning, having been found deeply unconscious; it was only the vigilance of an ambulance driver called to the scene, who drew attention to

Aortic stenosis
may lead to exertional syncope of a similar type.

H. J. Heimlich's own critical appraisal of the method appeared in *BMJ* **286** 1350 (1983). It is worth reading and warns that the back-slap effectively causes a bolus to impact further, except in inverted children.

Hyperventilation is often associated with syncope and carpo-pedal spasm. Treat by rebreathing the air in a paper bag or give O_2/CO_2 mixture by mask. Common in unstable teenagers.

the fact of the stove's being completely empty, that revealed the cause of the nearly fatal accident.

Miscellaneous

Cerebral tumours previously unsuspected may suddenly become swollen or bleed and produce alarming and dramatic symptoms difficult to diagnose.

Diabetes   Hypoglycaemic coma or keto-acidosis. Lactic-acidosis.

Uraemia   May be precipitated by infection in borderline renal failure.

Hyperpyrexia   Can develop in children rapidly in hot weather, especially in victims of cystic fibrosis.

Malaria (q.v.) — *Plasmodium falciparum.*

Sickle-cell crisis

Hypertensive encephalopathy

Wernicke's encephalopathy (vitamin B_1 deficiency) in alcoholics and the chronically starved.

Inanition/dehydration syndromes in the poor and neglected.

Metabolic disorders:

Myxoedema coma

Acute hypopituitarism, e.g. pituitary thrombosis following post-partum haemorrhage (pituitary apoplexy).

Acute adreno-cortical insufficiency can be due to incautious corticosteroid withdrawal, and acute stress (e.g. RTA) in a patient with undiagnosed adreno-cortical insufficiency (borderline Addison's Disease).

Waterhouse—Friederichsen syndrome   (See below.)

Acute septicaemic conditions: Influenzal myocardiopathy; meningococcal septicaemia leading to acute adrenal bleeding (Waterhouse—Friederichsen syndrome), which can also happen with other bacterial septicaemias.

Food poisoning (q.v.)   Various bacterial and viral food-poisonings can produce collapse, especially when accompanied in the elderly by severe vomiting and purging.

Toxic shock   Recent reports (e.g. *BMJ* **281** 1161 (1980)) of shock due to staphylococcal toxaemia in association with the retention of vaginal tampons have attracted a lot of attention. This is clearly a diagnosis to bear in mind but is relatively rare. Toxaemia from other forms of sepsis, local or general, is a first consideration especially in diabetics or people debilitated by disease or malnutrition. Septicaemia can arise secondarily to a suppurative tenosynovitis in the hand (e.g.) or without any apparent local source of infection. In A & E units group A haemolytic strep. is the common criminal. Treatment is with energetic combination of IV antibiotics, IV steroids, and IV plasma substitute. The possibility of using a vasopressor drug such as dopamine should not be ruled out when collapse is life-threatening.

Cf. **Tenosynovitis, Shock**.

Botulism is the most severe of this type and can be rapidly fatal. *Botulinum antitoxin is usually available at main hospital pharmacies.* Cf. **Food poisoning.**

Collapse can also be the presenting feature of Lassa, Ebola, and Marburg viraemias, malaria (*Plasmodium falciparum*), typhoid, rabies, diphtheria, poliomyelitis, smallpox (if it ever emerges again), and many other infections and zoonoses, especially in the debilitated or undernourished.

This vast and indigestible collection of causes of collapse and coma is included here for completeness's sake — but even so the collection remains incomplete. In a world of rapid travel and disturbed politics exotic diseases can turn up anywhere and every casualty officer needs to have them in the back of his mind. If he is lucky enough to have the experience of working in a country of the Third World he has to have them in the forefront of his mind.

What is important is that it should be made clear that primary care for every variety of collapse or coma consists in establishment of an airway and its maintenance; the supply of oxygen; supportive treatment, e.g. IV infusion, starting with glucose-saline (sodium chloride 0.18%, dextrose 4%) and the maintenance of the heart's action. Gradually the underlying condition may become manifest as more clinical and circumstantial evidence is accumulated, but generally more specific treatment is the responsibility of the receiving physician. However, the alert casualty officer who has suspected botulism because of the history, prostration, and muscular weakness of his patient, and has identified the nearest source of antitoxin before transferring his patient, will have deserved praise and gratitude which he is most unlikely ever to receive. He will however derive some encouragement from knowing that his care and diligence may have saved a life.

In subtropical and tropical climates it must be remembered that **poisonous bites** (especially in children) may present as collapse or coma. This applies especially to bites of cobra, viper, coral snake, or krait snakes in India, and of the black mamba in North and East Africa. In India a polyvalent antivenene against the four

common snakes is available and should be given as early as possible after injury. See **Bites**.

Scorpion and centipede stings can be fatal, especially in children, and are treated symptomatically. Jelly-fish poisoning likewise.

It must never be forgotten that wasp and bee stings (q.v.) can produce anphylactic reactions in hypersensitive people. These can be fatal in a few moments and anyone who has seen one will never forget it.

CONFUSION, ACUTE

Hypoglycaemia (q.v.)   IV Glucose 25–50 g; see **Diabetic emergencies**, p. 44a

Epilepsy   IV diazepam 5–10 mg

Toxaemia   Diagnosis and general supportive treatment, especially restoration of circulating blood-volume and blood-pressure*

Senile degeneration   Mild sedation and admission

Emotional stress   Elucidation before sedation if possible

CONVULSIONS, FEBRILE

Common in children and no different from any other epileptic attack. If there is status epilepticus protect the airway and give IV, IM, or rectal diazepam (given undiluted in usual parenteral dosage) and admit at once. Strong reassurance of staff and parents alike is the essence of the matter. In general, the treatment of the underlying pyrexia is not for the casualty officer.

CRAMP

Muscular cramp is little understood whether it occurs in the diaphragm (stitch), the abdominal muscles in swimming (in which case it may be fatal), or in the limbs. The last can occur after exertion or at night and can give severe pain. The commonplace remedy of anatomical stretching of the affected muscle is often effective but not always. In intractable cramp it is worth trying IV methocarbamol 100 mg/ml, 10 ml given over 10 minutes. See **Lumbago**.

The pain can be severe and is not to be belittled. A palpable

COMPUTERIZATION OF A & E RECORDS

This offers great hopes of advances in managing the vast and variable case-load of the accident service. If intelligently designed and applied with due sense of priorities it can provide an invaluable data base for clinical and epidemiological research.

Major psychiatric illness Specialist admission if necessary under the relevant section of the Mental Health Act (see **Psychiatric emergencies** and **Insane, certification of the**).

Hypoxia Especially in head injuries (q.v.).

*Dopamine: see p. 138a.

nodule may be felt and severe bruising may be followed by ischaemic necrosis and fibrosis. The condition is so obscure and ill-understood that rational treatment is unobtainable and the pathology at present not worth discussing.

CROUP

Or laryngeal stridor: a common seasonal concomitant of viral tracheo-bronchitis in children under five. Normally responds to
  Rectal diazepam, which can be a great help in panicky situations (some authorities disparage the use of sedation, others regard it as an important line of treatment): dose for age
  In severe cases IV Decadron 4 mg (dexmethasone)
  Steam — by tent
  Triominic — oral
  Cephradine — IM
98 per cent are cured in 24 hours by this regime; 2 per cent get laryngeal occlusion. Do you admit or not? If in doubt, admit, after discussion with a paediatrician if there is one. Very rarely intubation may be required or even laryngostomy (q.v.). This is mercifully rarely needed except in cases of diphtheria or injury.

Croup can be the presenting symptom of acute epiglottitis, which calls for early specialist attention especially in children. Occasionally urgent laryngostomy may be needed if there is acute anoxia. Do not hesitate.

CRYSTAL SYNOVITIS (gout and pseudo-gout)

This is a common cause of acute arthrosis (not only in the first metatarsophalangeal joint) in adults of all ages, often presenting as traumatic in origin. Sometimes a history of minor trauma is genuine but often it has been added by the patient as a rationalization of his pain and does not stand up to questioning.

Gout is the commoner cause and is identified by the following signs:
  1 Pain in the joint unrelieved by rest
  2 Local redness and swelling, with heat, but without fever
  3 Absence of history of trauma proportionate to the symptoms
  4 Raised serum uric acid (inconstant and unreliable)

Other causes of laryngeal stridor are excellently reviewed by Lissauer (*Hospital Update* **6** 821 (1980)), who gives a timely reminder of the role (forgotten in the UK) of diphtheria and retropharyngeal abscess.

Cf. **Respiratory emergencies, Foreign bodies.**

Acute epiglottitis (Haemophilus influenzae): temperature over 38 °C usually, child toxic; danger of cardiac arrest if intubation is attempted.

Joint aspiration is the only conclusive source of diagnostic accuracy (Hardy and Nation, *Archives of Emergency Medicine* **2** 89 (1984)).

    5   Response within 48 hours at most to colchicine or phenyl-
      butazone

Whether any further treatment beyond the immediate one is
required is a matter for a physician to decide.

The other cause of crystal synovitis is calcium pyrophosphate
deposition (pseudo-gout), which is diagnosed in acute cases by
identification of the crystals by polaroscopy and in chronic cases
by fluffy calcification in periarticular soft tissue. It is a good deal
less responsive to all forms of treatment than the genuine variety,
but phenylbutazone helps a bit. Uric acid crystals identify gout,
beyond any cavil. Juvenile gout is unusual but does occur.

DENTAL, MISCELLANEOUS

Bleeding extraction site

1   Get the patient to bite on a dental roll
2   Give IM penicillin and cloxacillin
3   IV tranexamic acid (Cyclokapron) 5 ml slowly, followed by
    oral amino-caproic acid (Epsikapron) 3 g twice a day or
    tranexamic acid 5 g three times a day for 48 hours. Topical
    application also helps. Cf. **Epistaxis**.
4   Pack with Gel-foam or similar material
5   If no good, suture with 3/0 Dexon
6   Refer to dentist in the morning
7   If brave, try ringing a dentist at home
8   Remember blood dyscrasia — rare, but not to be overlooked

Toothache

Dental abscess   IM penicillin and cloxacillin followed by oral
magnapen generally gives relief of pain in 10—12 hours. It is a
necessary precursor of dental treatment anyhow. Let the dentist
take over on Monday.

Dental caries   A painful cavity can be packed with a cotton-wool
pledget soaked in oil of cloves.

Dental fractures can be sealed with various applications. I use
Cavit or Kalzinol.

Displaced teeth

Should be firmly replaced digitally and splinted with double-folded aluminium cooking-foil, well moulded to the teeth. Stent is better still if you can get it.

DIABETIC EMERGENCIES

Hypoglycaemia

Common and urgent. Give IV glucose 50 g and observe for 2 hours at least. Further carbohydrate as circumstances suggest. Severe hypoglycaemia in a large adult may need a lot of IV glucose (e.g. 150 g). Admit children.

Hyperglycaemia

In general *not* an emergency. Never give insulin before admission. If circumstances compel you to do so *always* cover the insulin with IV glucose — 1 g to 1 unit. Always test for keto-acidosis. This and all other major diabetic emergencies need admission.

DIAZEPAM (Valium)

A reliable sedative to use in emergencies such as acute anxiety states and irritable or hyperkinetic head-injuries; appropriate for any age or type of patient except those in whom the insertion of a small IV cannula is impracticable. If it is administered according to the following simple rules, IV diazepam is a safe drug. If not, it can cause respiratory arrest with all the attendant complications.

Use IV diazepam in an emulsified vehicle to prevent phlebitis.

Give it *slowly* — the dose is judged by its effects. Start with 20 mg in the syringe and keep the patient talking as you press the plunger at the rate of not more than 5 mg every minute. When conversation dies away, and/or the eyelids cover half the pupil, cease the injection.

At every use of IV diazepam it is essential to have sucker and anaesthetic apparatus ready at the patient's head.

All preparations for immediate intubation must always be available.

Doxapram, 200 mg bolus IV (100 mg in children or small — under 44 kg (7 stone) — adults), is a selective respiratory stimulant

and to some extent a true reverser of diazepam narcosis. It is effective in 15—30 seconds and is not followed by relapse into unconsciousness. If it is given routinely after IV diazepam by way of the same IV cannula it removes all risk of respiratory complications and shortens the final recovery time.

An alternative method with a greater safety margin is to use a narcanalgesic cocktail. See **Narcanalgesia**.

DISLOCATIONS

Most simple dislocations can be dealt with simply and quickly on reception either with IV diazepam or narcanalgesia, or, failing success, with general anaesthesia. Notable dislocations:

Ankle   Usually in association with fracture, but simple dislocations can occur and are sometimes recurrent in ankles with collateral ligament injury.

Knee   May be anterior or posterior; either occludes the popliteal artery effectively and both need urgent reduction.

Patella   Easy to reduce with a knee extension jerk and gentle manipulation under general anaesthetic. Relaxation of muscles is important. The easier the reduction the more likely is recurrence of the dislocation. For this surgery is sometimes required. A POP cylinder gives safety and comfort.

Hip   Usually posterior and often associated with tibial plateau fracture in road traffic accidents. A concurrent fracture of the acetabular margin may produce instability of the reduction. This is best achieved in the anaesthetized patient, who should be lying on the floor. The manipulator stands over him holding the leg with 90° hip and knee flexion. Prolonged upwards traction precedes adduction and flexion of the hip. Open reduction is sometimes needed, but can generally be avoided if a muscle relaxant is given by the anaesthetist.

The severe pain of this dislocation demands early reduction.

In anterior and inferior dislocations the mechanism of dislocation is usually rotation of the falling or decelerating body about a fixed abducted leg (e.g. in a RTA the foot is caught under the seat and the body is thrown forward). Simple traction and

See 'Intravenous diazepam narcosis in the treatment of injuries, with Doxapram for recovery', Beesley, Bowden, Hardy, and Reynolds: *Resuscitation* **4** 259–63 (1976).

adduction is usually effective. If manipulation in the supine position does not work, try the semi-prone position.

Pelvis   Major pelvic dislocation at symphysis or sacro-iliac joint is generally part of a severe or multiple pelvic injury. Urinary complications must be looked for. Major blood-loss accompanies this injury in younger patients.

Phalanges or digits   One sharp longitudinal tweak; no anaesthetic, or a simple digital block. Unfortunately there are exceptions which may require open reduction:

1   Dorsal dislocation at the proximal interphalangeal joint with rupture of the extensor expansion. This also requires repair of the tendon to prevent development of the *boutonnière* deformity.

2   Volar (or sometimes other) dislocation at the distal interphalangeal joint with rupture of the capsule

Dislocated lunate   Reduce by extension and traction, followed by flexion. POP in flexion beginning with dorsal slab. Avascular necrosis may need later excision to avoid arthritis.

Elbow joint in children and adolescents   Helper applies traction at wrist (elbow at 90°); casualty officer applies firm pressure over olecranon with two thumbs; don't forget to check pulses and sensation. Unstable reduction suggests that there is a concomitant detachment of the medial epicondyle (with the elbow pulled into a valgus position), which has itself entered the joint, or allowed part of the relevant muscle group to do so. Forced extension and pronation or a faradic jerk to the common flexor muscles may free the obstruction. If not, open reduction is needed.

Pulled elbow   Presumably a distal subluxation of the radial head in young children. Diagnosis by history of traction injury to forearm followed by pain, pronation, and disuse; elicitation of mild local tenderness, limitation of elbow flexion and of supination of the forearm; misery. Radiography is in general *not* required. Treatment is by bimanual manipulation of the forearm. The elbow is held at right angles and a combined compression and pronation movement made quickly and forcibly. This should produce one cry from the child, a click at the elbow joint, and complete relief of symptoms and signs within 10 minutes. Result: happy child without disability, grateful parent, self-satisfied casualty officer.

3 Metacarpo-phalangeal dislocation of the thumb may need open reduction with division of the collateral ligament(s). These then need repair.

Shoulder   Reduce by Kocher's method, or the Hippocratic if that fails. If the patient has to wait for his anaesthetic, nurse him prone, with the dislocated arm hanging down. Very occasionally (with analgesics) spontaneous reduction occurs in this position. Always use Kocher's method first: it is relatively free of complications. After reduction, routine radiography is advisable to ensure reduction and no accompanying fracture. Kocher's method of reduction consists in four consecutive manoeuvres: (1) traction on the flexed elbow for 3—4 minutes, (2) external rotation of the arm, (3) adduction of the arm, (4) internal rotation of the arm.

The Hippocratic method consists in putting the unshod ipsilateral foot of the manipulator into the axilla and exerting steady traction on the arm with elbow extended so as to reduce the dislocation over the foot's fulcrum.

Fracture/dislocation of the shoulder is in general irreducible. Manipulation often makes the condition worse. Open reduction or excision of the head of the humerus may be required.

Jaw   This is always anterior when spontaneous (due to yawning, laughing, etc.). To replace, give IV diazepam — sufficient to produce calm and relaxation — then stand facing the seated patient; apply thumbs to mandible each side behind last molar; press downwards, while lifting angle of the jaw forwards and upwards with the fingers. Your reduction will be successful. Apply the standard first-aid bandage to prevent redislocation. Full anaesthesia is rarely needed.

DOSES, SOME COMMON

See Table 2, p. 48. All doses are approximate and need common-sense adjustment. Children are very tolerant of diazepam. Some people give 15 ml of syrup of ipecacuanha regardless of age.

TABLE 2   DOSES

Drug	Adult dose	Children's doses		
		0–2 years	3–4 years	5–10 years
Fortral inj. 30 mg/1 ml	30–90 mg	6 mg	10 mg	15 mg
Pethidine inj. 50 mg/1 ml	50–100 mg	—	—	—
Morphine inj. 10 mg/1 ml	10–15 mg	—	—	—
Cyclimorph = combined inj. of morphine 10–15 mg and antiemetic (cyclizine tartrate 50 mg)	10–15 mg	—	—	—
Omnopon 20 mg/1 ml	20 mg	—	—	—
Vallergan (trimeprazine)-forte 6 mg/1 ml	—	1–2 ml	3–4 ml	5–10 ml*
Junior aspirin = 75 mg aspirin	—	1–2 tablets	3–4 tablets	5–8 tablets
Junior Paynocil = 150 mg aspirin	—	½–2 tablets	½–2 tablets	2½–4 tablets
Diazepam inj. 10 mg/1 ml	10–100 mg	5 mg	5–10 mg	5–20 mg
tablets 2, 5, 10 and 25 mg	2–200 mg	—	—	—
syrup 2 mg/5 ml	5–50 ml	1 ml	2–5 ml	2–20 ml
Vallergan (trimeprazine) and atropine 10 ml (i.e. 60 mg) + 1.2 mg	—	1–2 ml	3–4 ml	5–10 ml
Stemetil inj. 12.5 mg/1 ml tablets 5 mg	5–25 mg	—	—	—
Ipecacuanha syrup	15 ml	2.5 ml	5 ml	10 ml

*This is the recommended dose. It is better to double it, wait 1¼ hours, then act.

DROWNING

Treat as you would asphyxia of any other kind: 1 Airway; 2 Intubation if necessary; 3 Tracheal suction; 4 Cardiac resuscitation; 5 Admission. Prolonged drowning produces electrolyte disturbance of some severity. Hypothermia (q.v.) may be a complication, especially in the aged. Swallowing of water occurs early; inhalation of water is generally small in amount and a terminal event.

DRUNKENNESS

The acutely drunk who are injured (and too often injure others) offer special problems: the diagnosis of their injuries is clouded by their drunkenness at every turn; they are difficult to handle and may assault staff, especially if they sense that staff are inexperienced or unsure of themselves; they are difficult to treat because they are irrational, uncooperative, and sometimes vomiting. The largest problem is to make sure whether they are just drunk, or drunk and ill, or drunk and injured. If you are in doubt, keep them under observation, preferably on the floor where they cannot fall any further. If you have no doubt that they are merely drunk, ask the police to cope. In the increasingly awkward medico-legal climate in Western countries routine blood-alcohol estimations (strictly confidential) give some protection to the casualty officer accused of negligence. They are also of value clinically in assessing the severity of concomitant head-injury.

Treatment

Plain drunks Stomach wash out (with sedation if violent). IV naloxone 0.4 mg at 10-minute intervals until sobriety is restored; this can make a restless and obstinate drunk submit to necessary treatment with a better grace. The violent drunk may need paraldehyde 10 ml IM into the gluteus maximus as a preparatory measure. There are no ill effects. Nikethamide no longer has any place.

Delirium tremens is a psychiatric emergency and to be treated as such, i.e. by certification, without sedation if possible. It is an acute withdrawal mania.

DRUG ADDICTION

A familiar problem in urban A & E units. The primary condition is not our responsibility but secondary effects may be, e.g. infection, acute psychiatric disorder, withdrawal syndromes, self-injury, and numerous associated psychopathic manifestations of manipulative kinds aimed at securing drug supplies. Encephalopathies occur in glue-sniffers as well as alcoholics. Disposal is a major problem.

Stomach wash out is designed to reduce risk of death, hasten recovery, and facilitate clinical care. It must *never* be seen as a punitive measure.

Sedation If this is necessary give IV or IM chlorpromazine, which can be supplemented with a high potency vitamin preparation such as IV parentrovite-forte. For the maniacally violent paraldehyde (up to 10 ml from a glass syringe) is safest, but has its well known disadvantages of discomfort to the patient and the persistence of its nauseating odour. IV or IM diazepam can also be useful, but is an additional, mild cerebral depressant.

DYSPHAGIA

A term used properly to describe difficult swallowing and improperly (but with equal frequency) to describe inability to swallow, or oesophageal occlusion (q.v.). True dysphagia is not really an A & E problem.

EAR, NOSE, AND THROAT EMERGENCIES

It has been said that there is only one ENT emergency, namely sudden perceptive deafness due to viral acoustic neuropathy, which requires urgent treatment with steroids, but some of the laryngeal crises and upper oesophageal calamities qualify as well. Cf. **Epistaxis, Asphyxia, Oesophageal occlusion, Viruses**.

EARS

Foreign bodies

Remove under direct vision if possible; general anaesthetic may be required.

Insects

Drown with warm olive oil or liquid paraffin before removal by forceps or syringe. The sensation of an insect fluttering on the ear-drum is wellnigh intolerable to some people.

Otitis externa

Gentle toilet and antibiotic eardrops (e.g. Otoseptil). If very painful and swollen, glycerine and ichthammol helps a lot; it is applied by a very gentle insertion of well soaked gauze 1 cm wide.

Avoid trauma; use oil; give antibiotics. Syringing in the presence of trauma or occlusion often does more harm than good. Use a loop or blunt hook, skill, and experience.

Otitis media

Common cause of disability in little children presenting as injuries. If this is severe and painful, and if there is a bulging eardrum, the need of myringotomy should be borne in mind. Antibiotic should be given in any case: amoxycillin or cephradine IM.

Pinna

Lacerations Careful approximation of skin edges with fine (5/0 or 6/0) sutures. It is important not to stitch through the cartilage as cauliflower ear may ensue if you do.

Haematoma Aspirate completely and apply pressure bandage to avoid cauliflower ear. The bandage (crêpe or Tubigrip) goes over a thick (2.5 cm), soft, sterile foam pad. Aspiration may have to be repeated several times, and a small instillation of Depo-Medrone (0.5 ml) may reduce the ultimate scarring. This is a difficult and time-consuming procedure, but untreated haematoma gives a horrible deformity. If haematoma involves the cartilage of the pinna itself it may need incision and drainage (fenestration) through the posterior surface. Antibiotic cover required.

ELBOW

Acute tennis-elbow

Frequently presents as an injury because of its acutely painful response to minor trauma. It is sometimes called non-traumatic epicondylitis of the elbow, but the commonest site of maximum symptoms is at the radial head or annular ligament.

Primary treatment with rest in a sling for 10 days with NSAID by mouth is often effective. If not, local injection of 1% lignocaine and 2 ml of a long-acting steroid (Depo-Medrone) is often effective. If this remedy is used first it is sometimes painful and relatively ineffective. Cf. **Tenosynovitis**.

Traumatic synovitis of the elbow

Commonly ensues after falls on the outstretched hand or directly on the elbow without X-ray evidence of injury to bone. It should always be treated by rest in a sling and patience. It may recover

When in doubt seek help from a plastic surgeon.

Subcutaneous retro-aural Depo-Medrone 2 ml gives good cover.

Complicating seroma needs referral.

Haematotympanum   Traumatic; usually associated with fracture of the floor of the tympanic cavity or of the petrous temporal bone itself. First aid: IM antibiotic and analgesia. Specialist referral advised.

A similar condition occurring medially is sometimes met and is called **golfer's elbow.**

very slowly (10 days to 4 weeks) and should never be hurried; physiotherapy is usually disappointing.

Hand and shoulder movements *must* be maintained throughout this tiresome period of recovery.

EMERGENCIES, MISCELLANEOUS

Ruptured plantaris tendon (so called)

Minor injury with surprisingly major symptoms. Characteristic history of sudden pain during active contraction; often bruising and tenderness just above maximum girth of calf. Support, and crutches if necessary. Painful for 10 days. What the pathology of the condition is is much disputed, but whether it is a torn vein (bruising may be severe), a muscular tear of gastrocnemius, or whatever, it is a recognizable syndrome with a label which makes the sufferer feel safer, and management more confident.

Traumatic arthrosis

Following sprains and blows, especially in the elderly, support and anti-inflammatory agents are harmless and helpful for a short period (5—10 days). Naproxen 250 mg in the morning, 500 mg at night.

The phenylbutazone series of NSAID has been withdrawn in the UK after more than 30 years because of the remote but well-known risk of inducing blood dyscrasia (1984).

Non-cardiac chest pain

may present with much anxiety. *Pectoral myalgia* (with superficial tenderness) is commoner; *Tietze's disease* (osteo-chondritis of one or more costo-chondral junctions) responds to local infiltration with bupivacaine and methylprednisolone acetate 40 mg.

ENDOCRINE EMERGENCIES

do not commonly present in A & E units but have to be borne in mind in differential diagnosis of **Collapse and coma** (q.v.). They include:

Thrombosed external pile

Simple incision under local anaethesia and evacuation of the clot gives relief. Admirably described in *Pye's surgical handicraft* (see p. 168), p. 446.

Cf. pseudo-coronary, p. 109a.

Thyroid crisis
Hyperthyroid cardiac failure
Myxoedema coma (elderly patients) especially in cold weather
Waterhouse—Friederichsen syndrome (supra-renal bleeding)
Phaeochromocytoma with paroxysmal hypertension
Pituitary coma
Pituitary apoplexy
Addison's disease

ENDOTRACHEAL INTUBATION

takes practice. You need to be able to do it instantly in every case in whom it is needed. If you cannot you will have to do a laryngostomy (q.v.). An anaesthetist is the best teacher, but an intubation model (e.g. the Vitalograph Intubation Trainer) is a good second best, preferably accompanied by the excellent Royal Free instruction sheets.

Prolonged intubation needs a clear plastic tube with the balloon moderately inflated. Measure the distance from incisors to cricoid and trim the tube so that it will lie comfortably with the connector at the lips and the balloon 2—3 cm below the vocal cords. If the tube is too long you may occlude the left main bronchus with consequent collapse of the lung.

Intubation is the paramount life-saving technique of the casualty officer. Positive-pressure ventilation can be maintained indefinitely using an ordinary anaesthetic apparatus or an Ambu-bag with an appropriate connector.

EPISTAXIS

Treat by:
1 Digital compression, for 10 minutes by the clock.
2 Sedation and rest for the agitated, in a position of comfort and drainage.
3 If these simple remedies are ineffective give IV tranexamic acid (Cyclokapron) 5 ml over 2 or 3 minutes, followed by oral amino-caproic acid powders (Epsikapron) 3 g twice daily for 3 days or tranexamic acid tablets 0.5 g, 2 twice daily for 3 days.
4 This remedy has a high success-rate, but occasionally the

Thyrotoxic crisis

Can present with a variety of symptoms, including eye-rolling, similar to the oculogyric crises of severe Parkinsonism (paralysis agitans or shaking palsy). The most urgent, however, is acute left ventricular heart-failure, which requires urgent sedation with diazepam in 10 mg IV aliquots every 10 minutes till the effect is achieved, and digitalization for acute fibrillation; flutter is controlled by IV propranolol 5−15 mg, hydrocortisone 100 mg; achieve cooling with fans and damp sheets. It is a grave emergency with irritability, confusion, delirium, hyperpyrexia, tachycardia, and diarrhoea in varying proportions (cf. mania, p. 129).

Myxoedema coma

Occurs in hypothyroid people in cold weather. They need ITU, triiodothyroxine 10 g IV, hydrocortisone, and standard treatment for **Hypothermia** (q.v.).

Adrenal crisis

Characterized by hypotension, shock, hypoglycaemia, stupor, and terminal coma. Usually seen in known cases of Addison's disease and meningococcal septicaemia (cf. p. 39). Treatment is IV glucose-saline and hydrocortisone 300 mg. Transfer to ITU after taking blood for baseline electrolyte-level.

Oculogyric crisis

A rare complication of Parkinsonism; there may be gross eye-elevation or conjugate deviation, combined with thought disorder. It can also occur, with phenothiazine medication especially, as a form of extra-pyramidal facial athetosis. Procyclidine 5−10 mg IV is effective. Not endocrine.

elderly arteriosclerotic or hypertensive needs packing. Use either bismuth oxide paste ribbon-gauze or ribbon-gauze soaked in liquid paraffin. Nasal packing *must* be done properly. If you do not know how, get someone who does to show you. It is built up, fold upon fold, with 2.5 cm ribbon-gauze until the cavity is *lightly* filled.

5 Very rarely an acute necrotizing rhinitis can bring about an epistaxis which proves fatal if untreated. This may well require ligation of both maxillary and both anterior ethmoidal arteries. I hope you never meet one.

EXTRA-MURAL AGENCIES

Ambulance staff

Ambulance staff give a high standard of first aid *en route* and are good assessors, on the whole, of seriousness of injuries. Patients with major or multiple injuries often need many hands for lifting, undressing, and splinting. Ambulance crews are experienced and willing helpers on arrival. They respond to gratitude like the rest of us. They also like to see X-rays and have things explained and evaluated if they are around when results come through. Accident departments have an obligation to maintain the interest and standards of training in ambulance staff, who respond enthusiastically to discussion, being asked for evidence, instruction, and general inclusion in the accident 'family'.

County Social Services Department

Difficult to get hold of but in theory always available somewhere to deal with social problems such as accommodation, financial aid, family difficulties. An originally willing and realistic service now slowly recovering from Seebohmization. They tend to get blamed for not producing the answers for the mass of psychopaths and sociopaths that are handed on to them. They are unfortunately human and have an uncertain role in society, being in some doubt whether their obligation is primarily to their 'clients' personally or to the social organization as a whole. Understanding and reciprocity pay a handsome dividend.

Less usual presentations as first symptom of
    Leukaemia
    Mononucleosis
    Thrombocytopenia
    Clotting and bleeding diseases (see p. 81)

Also responsible for certification of the insane. Contemporary difficulties arise from frequent changes of legislation regarding certification, place of safety regulations, and so on. All part of our bureaucratic mayhem. Treat symptomatically. (Cf. **Insane, certification of the**.)

Community nurses

Available in theory for dressings at home when distance or disability makes attendance at hospital difficult. Community nurses have a high standard of practice, but are pretty thin on the ground. Communication is often difficult. Contact direct if you can, by phoning the Nursing Supervisor at the Area Health Department, or via GP's surgery. Follow local usage.

Hospital medical social workers

Obtainable 5 days a week during office hours. Can sometimes help with no-fixed-address cases and other drop-outs: helping such people is often difficult, as accommodation is hard to find and their co-operation with plans made for their benefit difficult to obtain. Hospital medical social workers also have the unenviable duty of investigating the families of suspected battered babies (q.v.). They can often help, with the cost of their bus fares, injured people who have to travel long distances frequently for after-care. This can be a real problem for the impecunious who live far from hospital.

Police

Uniformly helpful but limited by legal considerations. Violent, disturbed, or inebriated characters who show signs of physical aggression and do not yield to the blandishments of charming nursing staff are often calmed by a police presence. Patients who are brought in and merely refuse treatment are best left to discharge themselves, if possible after signing a refusal form. Police are a cooperative and humane force much given to helping people and preventing trouble. It is almost too obvious to need stating that you need to help them if you want them to help you. Police and accident staff meet very often and have much to give each other. In general they cannot solve matrimonial disputes within the home, however violent.

Problems of confidentiality of information about patients sometimes cause misunderstanding. Frank and honest discussion generally resolves them as police are just as anxious as the rest of us to keep doctor—patient communications confidential.

Probation service

Sometimes willing to help in cases of matrimonial and family strife that have ended in violence, even where the opponents are not on probation. It is useful to know some of them personally in case probationers get into difficulties.

EYES

Chemosis, acute
See **Hay-fever**.

Corneal abrasions

Common in windy weather, gardeners, and people who put their heads out of carriage windows.

If blepharospasm is so severe as to prevent examination instil 2 drops of proparacaine 0.5% into the conjunctival sac, and then examine with a good light before and after instilling fluorescein drops, washed out subsequently with sterile water or saline. An abrasion shows a fixed green fluorescence.

If there is any corneal damage it should be treated with chloramphenicol eyedrops or ointment, and pad and bandage, and referred to the relevant eye hospital next day if not already healed.

Corneal foreign bodies

If superficial, these can be easily removed with the side of a 21 G needle after local anaesthesia with two proparacaine eyedrops. The corneal integrity should be tested with fluorescein afterwards, and treated as above if not intact.

Eye injuries

Refer to or send for expert advice from the relevant eye hospital. This includes all injuries to the lacrimal apparatus, tarsal plates, and full thickness injuries of the lids. Major injuries to the globe of the eye need early assessment and treatment, often at the same time as associated multiple injuries. Send every hyphaema to the specialist.

Swollen, contused eyes and ordinary oedematous black eyes need antiseptic or antibiotic eye-ointment until normal drainage is re-established.

X-ray the orbit for foreign bodies if there is any risk of these.

Send urgently for expert ophthalmic advice in cases of lacerations of the cornea or sclera, ocular concussion, or any other eye injury involving internal bleeding or external vitreous leakage, if they are combined with injuries preventing urgent transfer to a specialist unit.

Orbital 'blow-out' injury

in association with maxillary fracture involving the floor of the orbit can lead to entrapment of the inferior rectus. Warning signs are painful limitation of eye-elevation and infra-orbital sensory loss. This is an ophthalmic emergency requiring operative release of the entrapped muscle if permanent ophthalmoplegia is to be avoided.

FACIAL LACERATIONS

Primary suture by experienced and painstaking accident staff has much to offer. Plastic surgeons can offer little in addition unless the lacerations are associated with structural damage (see **Fractures**) or with skin loss needing emergency grafting operations.

The tiny, raised, often semi-lunar lacerations mady by glass spicules should be excised and sutured with the finest non-absorbable sutures if the patient's condition warrants it. If not stitched they cause disfigurement.

Look for damage to facial nerves, naso-lacrimal and parotid ducts.

Gravel-rash of the face needs 10 minutes scrubbing under GA.

FISH BONES

In the pharynx or oesophagus. The following pointers are useful in deciding if there is really a significant bone present:

What kind of fish? (e.g. cod bones are dangerous, kipper bones are not)

Orbital emphysema/pneumatocoele: see p. 63a.

Was the meal finished?
Is the pain localized?
Can the patient swallow, without grimacing, (a) water, (b) dry
    bread?

Do a careful inspection of the tonsils and pharynx, and an indirect
pharyngoscopy and laryngoscopy; fine bones often stick into the
pharyngeal or lingual tonsil. Give a large dose of ampicillin IM
and review in the morning. If there is severe pain and muscle
spasm, or a positive radiograph, inform the ENT department and
admit, especially so if there is pooling of saliva in the piriform
fossae. Removal of non-opaque bones demonstrated by Gastrografin
swallow may be required as an emergency. If you are in much
doubt refer for oesophagoscopy.

The same applies, of course, to other kinds of bones as well,
but these are generally radiopaque and so easier to assess.

FISH HOOKS

Have to be removed in the direction of entry because of the barb.
Inject 1 ml of lignocaine under the hook; cut the shank with
wire-cutters below the eye; seize the shank with a needle-holder,
press the point through the skin, and remove (Fig. 6).

An alternative method is to push the barb through the skin,
cut it off, and then withdraw the hook shank first.

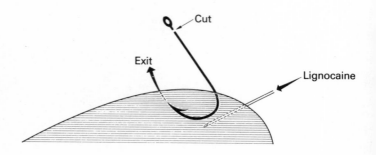

Fig. 6.

FOOD POISONING

Not a common A & E presentation, but the more dramatic forms may cause problems by their very unfamiliarity.

A careful history may help in differentiating between:

Botulism

Intoxication by *Clostridium botulinum* in tinned food, characterized by prostration and severe myasthenia after 8–24 hours of nausea and vomiting (cf. **Collapse and coma**);

Salmonella food poisoning

Commonly due to *Salmonella typhimurium* intoxication through rodent contamination of food; presents as a communal, febrile gastro-enteritis; and

Staphylococcal food poisoning

Due to staphylococcal infection of food-handlers – rapid onset and an afebrile course with early prostration.

Botulism may require assisted respiration because of respiratory neuromuscular block. It also needs 4000 U of polyvalent antitoxin IV as soon as a diagnosis is made, and every casualty officer should know where the antitoxin is kept – write it in the blank space opposite. Cf. **Collapse and coma**.

Primary care of salmonella and staphylococcal food poisoning lies simply in rehydration and then supportive treatment.

FOREIGN BODIES

Swallowed

Children will swallow anything, and what they can swallow they can generally pass. There can be no justification for serial radiography to 'follow the course of a foreign object through the digestive tract'. This exposes the child to a dose of radiation which gives him or her no imaginable benefit, however much it may reassure anxious parents or cowardly physicians.

Fish poisoning

An important article (Salmaso and others, *Lancet* ii 1124 (1980)) describes pelecypod-associated cholera in Sardinia. Other recent reports of fish-poisoning give important insight: Tatnall and others (*BMJ* **281** 984 (1980)) report a case of cigatuera poisoning. A fish bought in Antigua and salted was eaten later and transmitted the characteristic anticholinesterase toxin of the dinoflagellates eaten by the fish. Paralytic shellfish poisoning is due to a different dinoflagellate toxin (saxitoxin). Pralidoxime is probably of value in treatment, which is otherwise supportive. Atropine is useful for symptomatic relief.

A useful review article in the *BMJ* **281** 890 (1980)) surveys the whole subject of fish-poisoning. Several scombroid fish, if not eaten fresh, produce histamine and a related scombro-toxin, resulting in flushing, headache, and urticaria. Mackerel and Moroccan sardines have been implicated as well as tuna and bonito.

e.g. (West Midland Region)

Botulinum ABC antitoxin: Selly Oak Hospital, Birmingham, Emergency Drug Room (14), (021) 472 5313.

Acute infective diarrhoeas

Can present as emergencies. Prolonged diarrhoea with fever, bloody stools, or toxaemic collapse demand investigation. First treatment is the same for all: IV rehydration. Admit. Salmonella, shigella, campylobacter, *Escherichia coli*, vibrio, amoebiasis (p. 97a), and giardiasis are among the infective organisms and conditions. Physicians will sort them out.

The routine is:

1 Tell the parents of the almost infinite adaptability of the infant gut.

2 Explain that radiography neither cures, nor alleviates, nor determines a course of action.

3 Require them to give a normal or bulky diet and to examine the stools carefully for 5 days following the ingestion, or until the object is passed: expected time of arrival is from 48 hours to 5 days.

4 Abdominal pain should be reported to the GP, who must of course be kept in the picture.

5 If the object is not passed after 5 days it is justifiable to take a single straight film of the abdomen to give some idea of the progress of the journey. If the object is seen in the colon further patience is all that is needed. If it is still in the stomach, surgical opinion should be sought. Aperients are in all circumstances contra-indicated, but lubricant (liquid paraffin) is permissible for the constipated.

The following are among the objects which have, to my personal knowledge, been safely swallowed and ultimately reclaimed by natural means:

Open safety-pin	Tintacks
Teddy bear's eye with a 1½-inch wire hook	Drawing pins
Coins of all denominations	Hair-grips
Screws	Glass fragments

The teddy bear's eye was treated with the old-fashioned cotton-wool sandwich and arrived at its destination beautifully cocooned in that admirable protective.

Inhaled

These present an entirely different problem and should always be relentlessly localized and treated as a paediatric emergency *even in the absence of symptoms*. They can cause acute respiratory distress if impacted in the larynx, and can necessitate laryngotomy. They can lodge in one of the paralaryngeal recesses and cause chronic inflammation, and eventually acute laryngeal obstruction. They can get past the larynx and cause segmental collapse of the lungs, bronchiectasis, lung abscess, and all the associated sequelae. Routine: neck and chest X-ray and admit.

Rectal foreign bodies Almost invariably the input of sexual deviation. They need general anaesthesia and an experienced rectal surgeon as some of them do inevitable damage as they make their invariably dramatic exit.

Vaginal foreign bodies Commonly a forgotten tampon, but others are sometimes inserted by sexual deviants. Removal is usually simple. Self-abortion by (e.g.) knitting needle may penetrate the posterior fornix with very serious results.

Inhalation of a foreign body is always associated with some cough or choking effect.

Laryngeal impaction

It is rare for this condition to be seen in an accident unit as it has generally been dealt with by an enterprising bystander or proved fatal. If it occurs in small children it is best to pick the child up by the heels and give a smart blow between the shoulder blades. For practical reasons a different approach in large children and adults is needed. The Heimlich manoeuvre consists in delivering a sharp punch to the epigastrium in the midline so that an explosive force is built up in the chest sufficient to clear the larynx of obstruction. (See *BMJ* i 855 (1976).) Cf. **Asphyxia**.

Nasal

Often known or suspected by parents but may present in small children simply as unilateral, foul, nasal discharge.

Removal with nasal forceps, blunt hook, or loop after adequate suction is simple under GA, or in the conscious child, if properly held by experienced nursing staff.

FRACTURES

Diagnosis

Skill in reading X-ray films is partly the fruit of experience, but still more of being careful to the point of being pernickety. If you are in doubt about a particular bone, trace the outline of each of its constituent parts with a pencil point. You won't miss many fractures. If in doubt about odd bits of bone, X-ray both sides. Ultimately clinical symptoms and signs are the determining factor, rather than radiological appearances, which are at best distorted shadows.

Treatment

It is the job of the accident department to give primary treatment of adequate scope and sufficient skill to all fractures. If you are not happy about any particular case and a senior casualty officer is not available, get the advice of an orthopaedic specialist. It is impossible to overemphasize the importance of specialist orthopaedic care, from the beginning, for all cases of bone and joint injury. It is clearly impossible for an orthopaedic specialist to

Rhinolith is a rare outcome; being generally adherent it needs surgical removal.

deal personally with every such case accepted in the accident
department, but each remains primarily an orthopaedic responsibility. It is our job to treat all cases according to orthopaedic
opinion and advice so that they arrive at the ward or fracture
clinic appropriately sorted, supported, and comforted. Many
fractures can be definitively treated at first attendance and
referred to the fracture clinic for specialist supervision. A few
require specialist treatment from the start. *All* significant fractures
must be referred for specialist consideration at some stage.

It is clearly impossible to give a survey of fractures and dislocations (q.v.) in a short handbook, but one or two guidelines and
local variations may be of service. Orthopaedic practice is subject
to a large measure of personal variation, so you will need to alter
these observations to suit your locality. If your department
serves several orthopaedic specialists you will probably find that
they have different requirements for the same injuries.

Maxillo-facial injuries

Mandible   Undisplaced unilateral fractures are generally untreated
or given a supporting bandage. Displaced bilateral fractures need
wiring. If in doubt contact the orthopaedic registrar or the
maxillo-facial or oral specialist.

Nasal septum when grossly displaced can be reduced at once by
means of bimanual manipulation using a finger or small artery
forceps guarded by rubber tubing within the nasal cavity, or
referred to an ENT specialist for assessment on about the fifth
day (after swelling has subsided). Reduction advised for cosmetic
reasons, and to avoid embarrassed airway. Primary reduction of
the grossly deviated nose is far the most effective and needs
general anaesthetic or IV narcosis. Nasal packing with appropriate
ribbon gauze is usually needed in one or both nostrils. Packing is
built downwards from the apex of the nasal arch — the opposite
of the procedure for epistaxis. It also has to be firmer as its first
function is to maintain the reconstructed nasal architecture.

Nasal bones when fractured and much depressed can generally be
effectively manipulated in a similar manner. Subsequent ENT or
cosmetic surgery may be required, however.

Zygoma   Fracture common and not important. X-ray advised if

Careful X-ray studies to localize fractures in relation to dental roots are important. If you are in doubt get an orthopantogram done. It doesn't miss a trick.

A MacDonald's elevator is a useful tool for these. Meticulous care gives very good results. External splintage is used by some, but seems to confer little benefit and much discomfort.

in doubt. Elevation may be necessary for cosmetic reasons: this is generally done at a maxillo-facial unit or in an oral department. Severe zygomatic distortion can interfere with occlusion and chewing.

Other facial bones   Especially maxilla and floor of orbit. X-rays difficult to assess. Ask for 'OM 30' view to display the malar process of the maxilla. Fracture of the floor of the orbit may lead to herniation of orbital contents with consequent opthalmoplegia. See **Eyes**. Depressed malar fracture is very disfiguring and needs expert handling in its reduction and maintenance. Specialist assessment should be sought in case operative reduction is required.

Extensive facial fractures (varieties of 'dish-face') need internal reduction and complicated external splintage. A highly specialized job. It is rare for any of these to be in need of urgent reduction, and elective reference to the maxillo-facial specialist after consultation is generally enough. Some ophthalmologists would except the orbital blow-out but some not.

Neck

Fractures of cervical spine are important and, as far as atlas and axis are concerned, can be difficult to spot. They include the majority of unstable spinal injuries. When in doubt apply a soft collar firmly and get expert advice. If unstable fracture is suspected or certain, stabilize the head with sandbags or cushions. If the radiographer can get the patient to flex the cervical spine for X-ray purposes there is not likely to be a significant fracture. Any attempt at flexion where an unstable fracture is suspected is absolutely contra-indicated.

Major neurological involvement is generally obvious. In paraplegic or tetraplegic cases, fix the head with sandbags and send for help. It is essential to look at a lateral radiograph of the neck before permitting any movement for further views. It is not rare to see tetraplegic patients after injury with no evidence of fracture on X-ray. The dynamic situation in the cervical spine is very different from the static one after the disruptive forces have ceased.

Surgical emphysema of the orbit or *orbital pneumatocoele* is not an uncommon complication of maxillo-facial fractures involving the air-sinuses or nasal cavity. Pneumatocoele is characterized by sudden orbital inflation, usually on blowing the nose. Treatment is expectant, with antibiotic cover. Cf. Day and Engelhard, *BMJ* **281** 984 (1980).

Le Fort classification of maxillo-facial fractures is traditional but most oral surgeons prefer 'upper, middle, or lower third' divisions.

Obliques may help to show facetal fractures or displacements.

Every unconscious head-injury should have a 'single shot' lateral X-ray of the cervical spine before manipulation of the neck (e.g. for intubation or undressing).

Spine

Flexion injuries common. Antero-posterior wedging is the commoner, but look for lateral wedging too. Fractures of transverse and spinous processes are easily missed. They can be very painful but are not important. Unstable fractures, e.g. of the pedicles or articular processes, need handling with great respect, as described in standard texts (see **Bibliography**). Careful neurological examination essential. Uncomplicated fractures: rest till painless.

Ribs

Radiography not needed unless there is a clinical indication. Some patients demand strapping: few need it. A sling on the relevant side gives comfort. Local bupivacaine 0.5% (Marcain) gives effective relief. Pain lasts about 3 weeks untreated (see Fig. 7).

Stove-in chest and depressed sternal fractures are another story: the first is a class 1 emergency, the second needs morphine or pethidine. See **Chest injuries**.

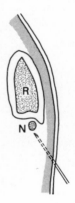

Fig. 7. Treatment of a painful rib (R). Inject about 5 ml 1% lignocaine or 0.5% bupivacaine around the intercostal nerve (N) at least 3 in proximal to the site of the fracture or injury. Repeat as necessary. **Intercostal nerve block.**

Flail chest, where many ribs and the clavicle are broken, is a class 1 emergency too: it can be rapidly fatal. Fixation of the flail segment is essential and life-saving. Endotracheal intubation with positive pressure ventilation will tide over the emergency until full-scale specialist treatment can be achieved.

Pelvis

In young people generally a severe injury due to major car or horse accidents. There is often visceral involvement and admission is required. Blood loss is severe and replacement is usually needed. Urine must always be tested for blood and a catheter passed only if directed by the surgeon responsible. Cf. **Uro-genital emergencies**.

In the elderly, fracture of the pelvis is usually due to a fall and is very common. Either the ischial or pubic ramus is involved: sometimes both. Treatment consists of 10 days in bed, with leg exercises, and if the patient has good support at home or in Part IV accommodation, admission is not required. This is a minor injury as a rule.

Upper limb

1 **Colles's fracture**   Looks easy to reduce and is difficult to reduce well. The Charnley method of reduction is gentle and satisfactory for many cases but is sometimes ineffective, in which case a more aggressive approach is required. The essential Charnley manipulations are traction, extension, and then flexion at the site of the fracture. When more force is required, traction, flexion, and ulnar deviation using knee or iliac crest as a fulcrum are the basic manipulations. Tactile comparison of the two wrists is the best criterion of the effectiveness of reduction. POP back-slab with the hand in *moderate* flexion and *moderate* ulnar deviation avoids subsequent displacement. The forearm must be in pronation during reduction and plastering.

Radial slab in early treatment of Colles's fracture is generally used so as to avoid impairment of circulation with consequent Volkmann's ischaemic contracture.

2 **Smith's fracture**   The reverse of Colles's fracture, i.e. fracture of distal end of radius with anterior displacement of distal fragment. This can be difficult to reduce and it is important to

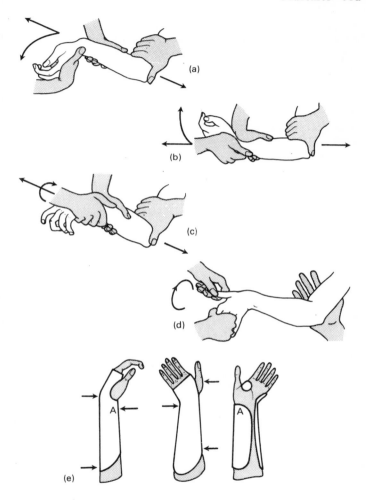

Fig. 7a. Detailed diagram of Charnley's method of manipulation of the commonest fracture. (a) Traction and disimpaction. (b) Two-handed flexion of wrist and fracture. (c) Pronation and moulding. (d) Plastering position: note exclusion of little finger. (e) Application of radial slab — well moulded at A.

reduce it fully to avoid disability afterwards. Don't forget that it requires above-elbow POP with the forearm in full supination to avoid redisplacement. Admit overnight to ensure supervision of the circulation.

3 **Barton's fracture** is similar to a Smith's fracture but only involves the anterior part of the base of the radius and is therefore inherently unstable. It is not difficult to reduce but requires to be put up in full supination with extension of the wrist in above-elbow POP. The carpus dislocates anteriorly with the anterior fragment. See Fig. 8.

Some people prefer to do ORIF as a routine using a small anterior T-plate (Ellis's) specifically designed for the purpose.

It is not all that common and so worth discussing with an orthopod when it arises.

4 **Fractures of radius and ulna** These are difficult and very important to deal with well. Manipulation should be attempted using an image-intensifier, and full plaster applied from the axilla to the heads of the metacarpals. Traditionally, fractures of the upper third are plastered in full supination, of the middle

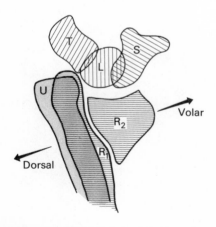

Fig. 8. Barton's fracture/dislocation of the wrist. U ulna; R1 shaft of radius; R2 anterior distal fragment; T triquetrum; L lunate; S scaphoid.

third in the mid-position, and of the lower third in full pronation. It has been suggested that re-angulation of fractures of the lowest sixth of the radius and ulna is best avoided if plastered in full supination. (Experience supports this method, but a clinical trial would settle any doubt.) Prolonged traction (2 or 3 minutes) enables you to 'hitch' the fractures; then manipulation is needed to improve their relationship; flexed fractures in supination and extended fractures in pronation (referable to the action of brachio-radialis). You can only really learn the 'feel' of these fractures by assiduous apprenticeship.

5 **'Juvenile Colles's fracture'**   This is not strictly a fracture but a posterior displacement of the radial epiphysis in patients under the age of 20 years (but usually under 15). There is generally a dorsal flake of bone broken from the radial metaphysis. It can be difficult to manipulate satisfactorily. Prolonged traction followed by forced flexion of the wrist first and then very firm digital pressure over the dorsum of the epiphysis is required. Below-elbow POP is used, with the forearm in strongly applied pronation. This is a stable fracture after full reduction.

6 **'Juvenile Smith's fracture'**   Similar to 4; to be treated as 2.

7 **Elbow**   If not displaced, too painful, or interfering with blood or nerve supply, can be left to be sorted out in the morning. Sling and analgesics will give relief.

Displaced fractures and dislocations (q.v.) about the elbow are tricky and need expert handling. This applies especially to fractures involving the epicondyles in children. Always check the radial pulse and hand movements.

It is very easy to miss fractures of the radial head or neck when looking at films. Look at every radial head with this specifically in mind *if there is local tenderness* of elbow.

8 **Humerus**   Unless grossly displaced or compound, collar and cuff. This does not apply to supra-condylar fractures (see 7). Different authorities recommend a variety of plasters for fractures of the humeral shaft. Verify local preferences.

9 **Clavicle**   Sometimes routinely treated with figure-of-eight bandage and *full* arm sling, sometimes by sling alone. Often a minor injury in children. (Obviously an occasion for a prospective clinical trial.) I think figure-of-eight with no sling is best.

The U-slab for fractured shaft of the humerus is applied over stockinette from the acromion to half-way up the inner side of the arm. The forearm is kept close to the upper abdomen during application. The whole is kept in position by enclosing the elbow at a right angle in plaster applied over plenty of wood-wool. It is very good if properly applied and very bad if not.

10 **Scapula**   Unless so gross as to require open reduction, full arm sling. Caused, generally, by direct violence.

11 **Carpal fractures**   Scaphoid fractures are commonest and most important because of the risk of avascular necrosis of the proximal fragment when the fracture is at the waist. Flexion and extension radiographs using a curved cassette increase your diagnostic score at first attendance. But if no fracture is seen on X-ray and local clinical signs are convincing, apply a scaphoid POP (thumb opposed to the middle finger and proximal phalanx included), re-X-ray out of POP after 3 weeks, and act accordingly. Fractures of the upper pole rarely cause problems and require symptomatic treatment only. Be sure never to miss fracture-dislocations such as the trans-scaphoid perilunate fracture-dislocation. It is agonizingly painful and easy to correct early. Not so later.

Others   After a few years in an accident department there will be no carpal bone you have not seen fractured. Mostly they require symptomatic treatment with POP for 3 or 4 weeks. Subsequent arthritic complications may require arthrodesis or removal of a markedly deformed carpal bone. Lunates normally, but not always, dislocate before they break.

12 **Stress fractures**   E.g. shaft of 2nd metatarsal. Gradual onset of pain with no early X-ray changes. May not be spotted until the excess callus of an unstable fracture is visible. Immobilization for 6 weeks may be needed. Can also occur in an athlete's tibia.

13 **Pathological fractures**   May be secondary to malignant deposits (e.g. carcinoma of breast, lung, or prostate); benign bone tumours such as chondroma (finger or toe); solitary cyst or fibrous dysplasia (metaphysis of long bone); Paget's disease in the elderly; brittle bones; senile or steroidal osteoporosis (vertebral most commonly). These are all common.

Von Recklinghausen's, myeloma, haemangioma, osteoclastoma, rickets, and scurvy can also cause fractures but are rare.

Primary malignant tumours rarely present as fractures.

Hand

It is rare for these to need extensive treatment or even reduction. Fractured metacarpals without gross displacement are best sup-

Stress fractures have been extensively reviewed in *Medisport* **2** No. 8 228 (1980) and **2** No. 9 262 (1980) by Michael Devas of Hastings, who has made them his special study. Further description here is out of place but the interested student is urged to read on.

ported by ring-strapping of the relevant fingers and kept slung until pain has diminished. Stable fractures of the hand and fingers, especially minor fractures involving finger joints, are best supported by elastic (Bedford type) supports, which allow greater mobility and give more comfort. Ring-strapping (adhesive garter strapping) is better where abduction injuries occur or when maintenance of length (e.g. in oblique or spiral metacarpal fractures) is required. Then active movement is important. In multiple fractures of metacarpals or other serious hand injury, support the whole hand with a volar slab in the position of function and admit for specialist care. Fractures of the phalanges are usually adequately dealt with by appropriate ring-strapping unless they are one aspect of a crush injury producing disorganization of hand-funtion. This is always a major orthopaedic problem which requires expert assessment and care. *Serious hand injury is an emergency and must always be treated as such.* If you are in doubt about the adequacy of simple support for a hand fracture get advice from a senior casualty officer or an orthopaedic specialist. Every A & E department should provide full facilities for emergency treatment of major or multiple hand injuries. They are of unparalleled importance to the patient, and the possibilities of salvage at primary operation are endless. They do, however, need experience, time, and the maximum of pains if good results are to be achieved. Details of treatment are out of place here, but at least strive to ensure that every such patient has the opportunity of the best possible care.

Special fractures of the hand *Bennett's fracture/dislocation of thumb* Anatomical reduction is essential. This may be stable in an abduction plaster with a small felt pad over the base of the first metacarpal. If not it needs internal fixation with a K-wire or screw (Fig. 9).

Shaft of first metacarpal An unstable fracture which may be stabilized in an abduction plaster, but needs an intra-medullary K-wire if not.

Second to fifth metacarpal shafts Unstable fractures may need internal fixation; stable fractures only need garter-strapping for 3 weeks and do well.

'Punch' fracture of fifth metacarpal neck Reduction makes this

If your own hospital does not offer an adequate hand-service, for the patient's sake refer him to another hospital that does. Too much depends on primary care for risks to be taken.

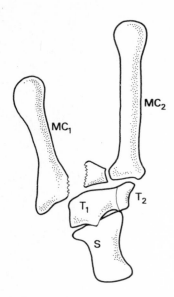

Fig. 9. Bennet's fracture/dislocation: an unstable injury. T_1 trapezium; T_2 trapezoid; S scaphoid; MC_1 and MC_2 first and second metacarpal.

fracture unstable and should not be attempted. Treat it by ring-strapping the two medial fingers for 3 weeks. The deformity will be permanent but slight, and function will be normal.

Phalangeal fractures   Find the stable position and splint with Zimmer padded aluminium splint. Some fractures are essentially unstable, notably oblique fractures involving the distal articular end (Fig. 10). These can sometimes be stabilized with a small K-wire.

The earliest mobilization of these fractures is vital. Various types of internal and external splintage are used to achieve this and expert advice should always be looked for.

Button-hole fractures of the terminal phalanx, in which the metaphyseal fragment is extruded and caught superficial to the nail fold, have to be carefully 'shoe-horned' in and under the nail fold and put in an extension splint (Zimmer type preferred).

Abduction fracture of the proximal phalanx of the little finger in

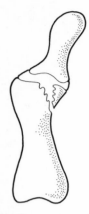

Fig. 10. Unstable fracture at the distal (or proximal) interphalan-geal joint.

children must always be reduced before ring-strapping. This is generally a juxta-epiphyseal green-stick fracture of the metaphysis of the proximal phalanx (Fig. 11). If left it leads to an unsightly and inconvenient deformity. See also Hand infections and injuries.

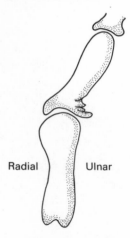

Radial   Ulnar

Fig. 11. Abduction fracture of proximal phalanx of minimus.

Mallet finger is often associated with a marginal fracture of the base of the terminal phalanx. Some people say that treatment should be symptomatic only as there is little residual disability. If a patient keenly wants treatment, a POP splint with the proximal interphalangeal joint flexed and the distal interphalangeal joint extended probably gives a 50 per cent recovery rate. Four weeks are sufficient (Fig. 12).

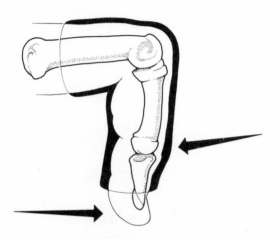

Fig. 12. Mallet-finger splint. Plaster bandage is applied to the extended finger and the whole is moulded to its final position and held while the plaster sets. Pressure is applied at the arrows.

Lower limb

1 **Femur, neck** Tie feet together as first aid if pain is severe. This is generally done by the ambulance crew.

2 **Femur, shaft** Thomas's splint with adhesive traction for first aid and before radiography. Give intravenous analgesic before moving.

3 **Knee** Above and below. Difficult fractures. Admit.

4 **Patella** With separation or comminution, admit. First aid:

The standard, plastic, mallet-finger splint can give approximately 85 per cent success rate if applied early and meticulously supervised. About six weeks is needed.

Femoral nerve block (p. 9a) should always be used before splintage.

splint in extension. 'Toffee in paper' fracture: POP cylinder. Do not be caught by congenital bipartite patella which is often (but not always) bilateral.

5 **Tibia and fibula (shaft)** Unstable fracture needing expert mending; a foam trough is better than an inflatable splint for first aid support and radiography. Manipulation and above-knee POP after orthopaedic consultation. Better reduction with less risk of angulation is achieved by using the vertical position with the knee flexed over the end of the trolley. The plaster should then be applied in three parts, first the leg while traction on the heel is maintained, then the foot, then the above-knee portion.

Compound fractures need careful wound toilet and closure with Dexon sutures as well as full antibiotic cover and tetanus protection before manipulation and plastering.

Remember that such fractures represent between 1 and 2 litres' blood loss. Admit.

6 **Ankle fractures** *Pott's* All degrees of Pott's fracture require POP but there is wide divergence of opinion about when internal fixation should be used. You will have to find out what the relevant orthopaedic specialist prefers and act accordingly.

Fibular Below inferior tibio-fibular joint, POP; above inferior tibio-fibular joint (undisplaced), double Tubigrip or POP if pain is persistent. If there is diastasis at the inferior tibio-fibular joint, internal fixation may be required.

Unstable ankle fractures and ligamentous injuries Except for unstable fractures of the malleoli, which need internal fixation, ligamentous injuries must always be demonstrated radiographically by means of stressed inversion/eversion X-rays under general anaesthesia or IV narcanalgesia (see **Narcanalgesia**). Rupture of the medial ligament of the ankle needs operative repair, and of the lateral ligament POP in a slightly everted and extended position of the ankle joint for 6 weeks (cf. **Ankle**). Some repair both, others repair neither. Rational investigation of the efficacy of these contradictory approaches seems to be called for.

7 **Foot** *Calcaneus* Common after jumping. Tender on lateral compression; painful and disabling. Severely swollen injuries benefit from high elevation. Tubigrip and crutches; ultra-sonics

Look for a fracture at the fibular neck in cases of inferior tibio-fibular diastatis. It is an indicator of Maisonneuve fracture. In this type the interosseous membrane itself may be split, so that the two bones open apart like a pair of tongs. Very much an ortho-paedic problem needing ORIF. Cf. **Ankle**, ruptured ligaments of.

for pain. Some people reduce all flattened fractures; others reduce none of them. Some recommend primary subtalar fusion for bad calcanean fractures. Transmitted violence may produce associated fractures of the tibial plateau, dorso-lumbar spine, and odontoid peg. If severe, admit for elevation and review.

Talus and other tarsal bones   Comparatively rare. Local tenderness diagnostic. Specify site of local signs to radiographer and radiologist. Orthopaedic advice needed in cases of fracture, especially of the talus and navicular (tarsal scaphoid).

Metatarsals   Common; especially the base of metatarsal 5 in acute inversion injuries. This only needs support except for the rare cases of gross displacement. Other metatarsal shafts (especially stress fracture of metatarsal 2 or 3 — of which the only radiological signs may be callus-formation) need only support and rest, unless much displaced. Varieties of transverse tarsal fractures and tarsal crush injuries occur rarely and need specialist assessment from the beginning.

Phalanges   Very common in industrial and domestic accidents. Usually satisfactorily treated by ring-strapping. In crush injuries with avulsion of nails or burst injuries, don't forget that these fractures are compound and need a full dose of intramuscular penicillin and cloxacillin, and tetanus protection. Oral antibiotics may be desirable as well. Exploration and reconstruction may be required as in crushed fingers (see **Hand infections and injuries**).

Non-fractures

Bipartite patella
Sinding—Larsen—Johansson disease of lower pole of patella
Looser's zones in osteo-malacia
Pseudo-fractures in Paget's
Osteochondritides
Accessory ossicles especially of talus and navicular
Pellegrini—Stieda syndrome (look it up!)
Tibial exostoses (may need excision for relief of symptoms)
Normal but partly ossified epiphyses in children are diagnosed as fractures with lamentable frequency: Reference to a textbook of anatomy sorts them out

Distal and mid-tarsal fracture-dislocations are very destructive injuries, not uncommon in RTA. Painstaking realignment and fixation are specialist problems.

Osteogenic sarcoma may present as a swollen and tender meta-physis. Characteristic X-ray appearances identify it. In the early stages it is sometimes referred for physiotherapy, lead-ing to tragic delay.

GAS GANGRENE

Recommendations sent out from the Good Hope Hospital, Sutton Coldfield (10 September 1974), are reproduced here, as they deal so admirably with this subject. With appropriate alterations they should apply in other regions.

1   The diagnosis of gas gangrene should be made on clinical grounds, when a patient has gangrene with muscle necrosis, is toxaemic, and has crepitation in the tissues due to gas formation.

2   Surgery for the condition should be undertaken after hyperbaric oxygen therapy has been given. The mortality of the disease is greatly decreased if the patients are first detoxicated, and therefore surgery undertaken when the pulse rates and temperatures are normal. Surgery after hyperbaric oxygen is safer, and often less radical than is otherwise necessary when operating on very toxic and ill patients.

3   The beneficial effects of antitoxin are in doubt and probably outweighed by contra-indication. Antitoxin is therefore rarely given especially when hyperbaric oxygen treatment is available. Penicillin G, 2 mega units 4-hourly, should however always be given.

4   *If gas gangrene is suspected, penicillin should be started, and arrangements made to transfer the patient to the Hyperbaric Oxygen Unit at Good Hope General Hospital, Sutton Coldfield* [or to the nearest centre with these facilities], *as soon as possible*. Surgery will be performed at Good Hope after the patient has had initial treatment with hyperbaric oxygen.

We believe that if the above measures are accepted, the mortality of gas gangrene could be greatly reduced.

GASTRO-INTESTINAL EMERGENCIES

Briefly reviewed as an aide-memoire:

Gastro-intestinal bleeds

Oesophageal varices
Bleeding peptic ulcers
Mallory—Weiss syndrome (bleeding after vomiting)

Haemangiomata
Meckel's diverticulum (containing ectopic gastric mucosa)
Carcinoma of the rectum more commonly than of the colon
(See **Collapse and coma, Haematological emergencies.**)

Food poisoning (q.v.)

Obstructive syndromes

Hypertrophic pyloric stenosis in infants Usually a male first-born with projectile vomiting and a palpable tumour while feeding (3—8 weeks old). Can occur in females and siblings.

Volvulus of loops of small intestine round or through adhesions, Meckel's diverticulum or whatnot.

Strangulated herniae Umbilical, intestinal, inguinal, femoral, and obturator.

Intussusception Intermittent vomiting and colic in bonny, 6—9 month-old boys. Palpable tumour far from universal.

Faecal impaction in elderly, debilitated patients presents from time to time with pelvic pain and pseudo-diarrhoea. Needs manual removal, but should never come to it.

Miscellaneous

Appendicitis
Gall-bladder disease
Pancreatis
Diverticulitis
Crohn's disease
Various neonatal atresias — not likely to present at A & E departments in developed countries
Acute rectal prolapse in infants needs gentle manipulative reduction, strapping of the buttocks and admission.
Gall-stone ileus The author has seen one case of cystine-stone ileus arising from a perinephric abscess which ulcerated into the jejunum; not a common event.
Abdominal trauma is a separate subject and is dealt with under **Abdominal injuries** (q.v.).

Neoplastic   disease   A   rare   cause   of   acute   gastro-intestinal
bleeding.

GPs

Friendly and close collaboration with local GPs is one of the foundation-stones of successful management of an accident unit.

If the GPs provide a conscientious service to the patients (which in my experience they do magnificently) the unit can get on with treating accidents and emergencies. As a *quid pro quo* the unit should be ready at all times to help with urgent diagnostic conundrums and can often provide an invaluable service in regular sessions for non-urgent minor surgery. Urgent minor surgery is our responsibility anyhow.

GRAFTS, SKIN

In accident work which is based on out-patient treatment only, there are three types of skin-graft which may be useful:

Pinch grafts

For small finger-repairs (excluding finger-tip ablations which are better dealt with by V—Y plasty (p. 85) or volar flaps). These grafts are done by raising a weal on the hairless skin of the forearm and slicing off the bleb with a scalpel blade. They are transferred to the denuded area after careful cleaning and debridement, fixed with 4 × 5/0 Dexon stitches, flattened, covered with a single layer of chlorhexidine tulle and a firm pressure bandage, and left in place for 5 days in a sling for elevation. Cover donor area with chlorhexidine tulle. After 5 days the grafts are inspected *very carefully* and lightly dressed as the situation demands. Once well 'taken' the graft should be kept well greased.

Split skin grafts

Can readily be taken from the donor area under local anaethesia. A simple technique with excellent illustrations was published in *Teach-in* (March 1973, pp. 187—92) by Roger Snook. It would be useful to reproduce it but too expensive. These grafts are useful for *backs* of hands and feet with extensive skin-deficit.

With experience small Thiersch grafts can be cut with a scalpel, but a Humby knife is simpler. They should be handled very carefully, stitched in place with 5/0 Dexon or 6/0 Prolene sutures and punctured with a scalpel after fixation so as to prevent haematoma or seroma. Firm dressing for 5 days is essential. Bactigras (chlorhexidine tulle) is the best first layer to avoid sticking.

Pedicle grafting

Useful for full thickness skin-loss from fingers. A pedicle is simply made in the subcoracoid fossa of the contralateral shoulder (which is generally hairless) and the graft applied to the deficiency on the hand. The whole is kept in place by strapping the hand to the chest with 3-inch Elastoplast, and the pedicle can be divided after 14—21 days. Careful follow-up is essential to keep at bay the inevitable minor infection beneath the neck of the graft where you make the flap into a tube. The technique is standardized and need not be described in detail (see **Bibliography**, Surgery, p. 168, *Fundamental techniques of plastic surgery*). Small pedicle grafts for fingers do not require Thiersch grafting to the donor area as primary suture will generally close the deficiency.

Wolfe (full thickness) grafts

Not as reliable as the above but may be attempted when a patient brings in the skin he has lost. The auto-graft should be defatted completely and with care, and the graft thoroughly washed in saline first and then cetrimide (0.5% solution). It needs to be fixed with numerous fine peripheral stitches and pierced to avoid haematoma. Pad and bandage provides compression and the whole should be left alone for 5 days at least. Oral antibiotics and elevation of the limb.

GUN-SHOT WOUNDS

In general gun-shot wounds with major damage occur in young people and need early blood-volume replacement.

Airgun

Remove subcutaneous pellets and leave deeper ones unless they are near vital structures — e.g. in the wrist. Pellets land in the orbit surprisingly commonly and often do little damage. These should be referred for an ophthalmic opinion, but can frequently be left alone.

·22 rifle

The same rules apply, with the proviso that the bullet must be

carefully localized by X-ray. Abdominal wounds need to be regarded with extreme suspicion, as visceral perforations can occur with a minimum of initial symptoms and signs, and no demonstrable track.

·410 shot-gun

Dangerous only at close range (a few feet). Although the explosive charge is small the muzzle-velocity is high enough to do considerable damage. For instance, of two injuries due to stumbling when holding a loaded ·410, one necessitated resection of the right lobe of the liver after ligature of the right hepatic artery, and the other resulted in death due to multiple perforations of the aortic bifurcation with massive haemato-peritoneum.

Twelve-bore shot-gun

A very much more powerful gun which can produce extensive damage at a range of several yards. Suicidal shooting through the mouth is immensely destructive. Stray pellets are a common finding in routine X-rays of rural patients and require no action. The sawn-off shot-gun is a familiar cause of destructive injuries at close range owing to the rapid spread of the charge. The injuries are unlikely to be the casualty officer's worry as they require definitive surgical treatment according to general anatomical and surgical principles. They are difficult cases to deal with owing to the wide extent of tissue damage and skin destruction. Urgent resuscitation and blood-transfusion are generally required.

Military missile systems (e.g. Armalite and other high velocity rifles)

These have their own characteristics and are dealt with in specialist literature (see **Bibliography**).

GYNAECOLOGICAL EMERGENCIES

Abortion (q.v.)

Ectopic pregnancy   Low abdominal pain, amenorrhoea; bleeding not always present.

Endometriosis   Can present as an acute apyrexial parametritis or

Explosive bullets

An increasing menace in criminal assaults. *BMJ* **284** 768 (1982).

Gun-shot and bomb-blast injuries (Belfast)

Roy, *Journal of the Royal Society of Medicine* **75** 542 (1982).

intestinal obstruction, or just as pain; often related to menstruation.

Ovarian cysts   Bleeding into or torsion of.

Salpingitis Febrile, acutely painful, and tender; localizing on internal examination; vaginal discharge not always present.

Vulvo-vaginitis can have a very acute onset and present at night or week-ends. Warm bicarbonate bathing and/or douching is a good first-aid measure.

Trichomonas, fungal and **monilial infections** should be eliminated by submitting appropriate specimens to the laboratory; treatment with polyvalent pessaries and creams can be begun if needed. Predisposing causes (diabetes, poor hygiene, malnutrition, and debility) should be excluded. Return to care of GP.

Cf. **Obstetric emergencies, Abscesses, Venereal disease.**

HAEMATOLOGICAL EMERGENCIES

See Table 3, p. 81.

HAND INFECTIONS AND INJURIES (cf. **Fractures, hand; Wrist and hand amputations**)

Many hand infections and injuries are minor and can be dealt with according to general principles. But there are special aspects because the hand is a miniaturized sensori-motor system of complex structure and is exposed to special types of trauma.

Crushed terminal phalanx

An important injury if there is a compound element. If there is fragmentation of the bone with displacement and soft tissue injury, exploration under digital nerve block is essential. Detached fragments should be removed and careful suture performed. Neglect of this can lead to sequestration of fragments and long delayed recovery. IM penicillin and cloxacillin is advisable. Tetanus protection is required. 5-day course of oral antibiotic after reconstruction of all compound injuries. Ampicillin/flucloxacillin type preferred in prophylaxis (e.g. Magnapen).

Subungual haematoma

See **Nails, digital**

(Cont. on p. 82)

Thalassaemia

A cheap and simple test for mass screening in susceptible populations described in *BMJ* **286** 1007 (1983) by Silvestrino and Bianco.

Augmentin probably preferable in 1984.

TABLE 3 HAEMATOLOGICAL EMERGENCIES

Class		Type	Cause	Treatment	Comments
Abnormal bleeding	Congenital	Afibrinogenaemia	Factor I deficit	Fresh frozen plasma (FFP) or fibrinogen	Rare: usually documented; serious after trauma
		Christmas disease	Factor IX deficit	IX concentrate or FFP	Generally known to haematologist
		Haemophilia	Factor VIII deficit	FFP or VIII concentrate or cryoprecipitate	Ditto; most carry cards
		von Willebrand's	Factor VIII abnormality	FFP or VIII concentrate or cryoprecipitate	Ditto; tranexamic acid (Cyclokapron) for minor episodes
	Iatrogenic	Warfarin	History of myocardial infarction or deep vein thrombosis	FFP or IV vitamin K (phytomenadione)	Common and not very severe
		Heparin	Ditto	Protamine sulphate;	Usually in hospital; rare nowadays, and severe
		Dindevan	Ditto	IV vitamin K analogue or FFP	
	Leukaemic	Thrombocytopenic or hepatogenic (interference with fibrinogen production)	Abnormal bloodcell production and marrowfailure	Fresh blood; platelet concentrate or FFP as emergency treatment	Early reference to specialist unit
	Thrombocytopenic (primary)	Immunogenic in children	Obscure	No emergency treatment usually required but platelets for mortal bleeding	Admission to specialist unit; steroids
		Various forms of thrombasthenia*	Platelet abnormality		
	Defibrinalization disease	Chronic bleeding disease	See Abnormal blood loss below	Fibrinogen or FFP	Primary condition needs sorting

	(cf. thalassaemia)	opathy, reduced oxygen tension and acidosis precipitates	same + antibiotics	ranean and some Indian genetic association; refer to specialist hospital as soon as possible
Abnormal blood-loss	Dysproteinaemia — Waldenström	Raised IgM	Plasmaphoresis	Refer to physician or haematologist
	Haemoglobinuria — Agglutinin abnormalities	Cold, March, paroxysmal (i.e. idiopathic)	Reassurance	Admission
	Bleeding from alimentary tract — Oesophageal varices	Hepatic fibrosis; congenital absence of spleen/portal vein	Transfusion of whole blood or packed cells; Sengstaken tube	Admit
	Peptic ulcer / Oesophagitis	Who knows?	Transfusion of whole blood	Admit
	Mallory–Weiss syndrome	Gastro-oesophageal bleeding after vomiting	Referral for diagnosis	Rare
	Meckel's diverticulum	Containing ectopic gastric mucosa	Ditto	Ditto
	Intestinal polyposis	—	Ditto	Ditto
	Intestinal haemangiomatosis	—	Ditto	Ditto
	Colonic angiectasia	—	Ditto	Ditto
	Malignant disease	God knows!	Ditto	Ditto
Miscellaneous	Abdominal emergency — Spontaneous rupture of the spleen	1 Infective mononucleosis (EB virus) 2 Malaria	Surgical after urgent blood replacement	1 Rare 2 Common in malarial countries

*See enlightening report by Briët and others, *BMJ* **281** 1039 (1980).

Burst finger

Common industrial accident. Prognosis often bad in view of
vascular damage and poor viability of tissues. Minimum of repairs;
POP volar slab until comfortable (not more than 7 days); sling;
high dose antibiotic. Refer to hand clinic or for specialist care if
required. Primary closure may be essential to obtain skin cover. If
the damaged portion is unable to be salvaged because of fragmentation, loss of tissue or deprivation of blood-supply, terminalization
of the next phalanx may be necessary, always using a volar flap.
Some sacrifice of length needs to be accepted to achieve this.

The safest and best splintage for all difficult hand-injuries is
the full-length (elbow to finger-tip) volar slab bandaged only to
the metacarpophalangeal joints. The forearm is pronated, wrist
extended, metacarpophalangeal joints at 90° of flexion and interphalangeal joints as nearly extended as possible. This last can be
modified if early movement is possible, insisted on, and supervised.
It is excellent for all dorsal injuries including comminuted fractures, extensor tendon repairs and severe skin-loss of the dorsum.
Dressings to fingers should be kept to a minimum and done
without if possible.

Early movement takes precedence always. An ugly finger that
moves is better than a beautiful finger that doesn't.

Lacerations

Provided nerve and tendon damage is absent and the blood supply
is good, careful suturing with monofilament nylon or polypropylene, 4, 5, or 6 grade, which gives the best scars.* In children *a lot
can be done with Steristrip*, because the skin is soft. This is both
kinder and more effective, when circumstances permit.

Digital nerve repair is generally accepted as a primary procedure
more likely to give quicker and more effective recovery of sensation than not. It is a simple but tedious procedure requiring very
careful dissection of the nerve-ends using an operating-loupe* and
uniting the epineurium of the digital nerve in two or more places
with the finest possible non-absorbable sutures; 8/0 or 10/0
polypropylene or virgin silk. The nerve should be correctly
rotated, handled gently and little, and supported for the first 10
days in a position of rest and laxity.

*The finer the better provided the grade is appropriate to the wound.

*Or operating microscope: this will be essential if repair of the digital artery is to be done at the same time.

Infections

Paronychia (whitlow), hyponychia, perionychia, pulp infection, and other deep infections of the fingers can be perfectly well dealt with under digital nerve block provided the local anaesthetic is injected into healthy tissue and concurrent antibiotic given in adequate dosage. The secret of success is adequate exposure, exploration, and insertion of a small wick to ensure drainage. Chlorhexidine tulle in the first instance is kinder than gauze, but should not be continued after the first dressing, as it makes the skin soggy; the minimum amount always to be used.

Chronic paronychia (fungal or monilial with secondary bacterial overtones) can dawdle on for months with painful episodes of secondary infection. Combined attack with local toilet of the nail-fold, oral antibiotics, and prolonged topical antifungal/monilial agents, e.g. 1% clotrimazole cream (Canesten). Sometimes surgery is required to clear out perionychial loculi, but not often. Topical applications for 3 weeks at least.

Palmar space infections and abscesses involving tendon sheaths

require the fullest exploration, drainage and after-care. Refer to a senior casualty officer or orthopaedic specialist. Sling and high-dose antibiotic is desirable, but immobilization should if possible be avoided because of the risk of stiffness. Irrigation with a solution of H_2O_2 after surgery gives a better and quicker resolution. There is *no* place for stab incisions and short cuts. If local anaesthetic is inadequate, general anaesthetic should be resorted to without hesitation to secure adequate access and treatment. For example an area of skin equivalent to the circumference of a deep pulp abscess should always be removed to ensure adequate drainage (see Fig. 13). IM penicillin and cloxacillin is the treatment of choice. IM clindamycin in cases of penicillin sensitivity or where anaerobic infection is suspected or proved. Careful visual exploration of every hand injury and infection is essential; detailed and meticulous surgery and intelligent and sympathetic aftercare are equally important. The treatment of pulp abscess by the traditional through-and-through stab incision, followed by a wick, is never to be considered. Where there is pus inside the tendon sheath adequate drainage is vitally important, early movement

Major infections need parenteral antibiotics 8-hourly for 24 hours or longer — e.g. penicillin G 600 mg + flucloxacillin 500 mg, or clindamycin if there is any suspicion of anaerobic infection — as well as proper drainage. After-care with frequent warm saline bathing and povidone-iodine ointment.

Other conditions affecting the fingers, especially the nail-fold, may mislead the unwary, e.g. granuloma annulare, mycosis, Boeck's sarcoid, erysipeloid.

Oral systemically-acting fungicides help, e.g. Griseofulvin 500 mg twice a day.

NEVER STITCH AN ABSCESS OF THE HAND, only exploratory incisions.

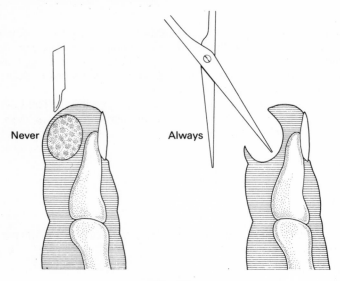

Fig. 13.

essential, and 4-hourly irrigation with topical antibiotic solution
recommended by some, using a fine catheter in the wound.
Experience validates this method and gives rapid and complete
suppression of sepsis and early return to full function. I use 2
strips of Yates drain of different lengths sutured into the finger
alongside the infected tendon; the longer at the proximal end of
the infected area and the shorter about half way along it. These
are left in place for 48 hours and used to convey the instilled
antibiotic (e.g. clindamycin solution 300 mg 4-hourly) to the
whole infected area (see Fig. 14). This does not delay healing and
produces dramatically successful resolution of a process all too
often ending in disaster. 48 hours is the limit. Don't forget to
use IMI antibiotics in all cases of deep finger injection. Although
pus may be present in a localized area, the localization may not
be complete. Where there is a concentration of vital structures
(tendon, bone, neurovascular bundle) in such a small space it is
not justifiable to take any risk whatever.

See also **Tendon repairs, hand.**

An alternative is to use 2 fine IV catheters (e.g. paediatric gauge 20), one halved in length, which are easier to introduce into the tendon-sheath but also easier to kink.

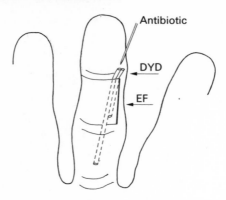

Fig. 14. Basic features of treatment of suppurative tenosynovitis. DYD double Yates' drain; EF exploratory flap.

Amputated finger tips

Thiersch thickness: dress with chlorhexidine tulle and allow to epithelialize.

Full thickness without loss of pulp (i.e. some dermal remnants): postage stamp Thiersch graft from hairless skin obtained under local anaesthetic with a scalpel.

Full thickness with pulp loss or exposure of terminal phalanx: V−Y plasty (see below) or Wolfe-graft. Rarely a pedicle-graft from the contralateral chest wall is needed. If you can do it a V−Y plasty is best.

V−Y plasty

The purpose of this simple and elegant operation is to provide two symmetrical sliding pedicle-grafts from the sides of the finger which can be brought up to unite over the denuded finger-tip (Fig. 15). It provides skin cover of the right type, carrying its own blood and nerve supply. The best description of it is in Kutler's original paper (*Journal of the American Medical Association* **133**, 29−30 (1974)). A similar V−Y advancement of volar skin can be used to make good oblique ablations of the whole pulp of a finger or thumb. It is easier to perform and works well in younger

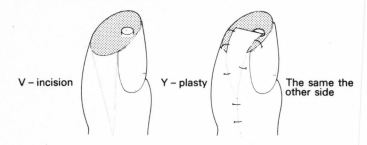

V – incision   Y – plasty   The same the other side

Fig. 15.

patients with soft skins, but has the disadvantage of leaving a volar scar. See Fig. 16.

Traumatic amputations

of fingers require careful terminalization of the remaining phalanges. Volar flaps to cover the finger-end should be used whenever

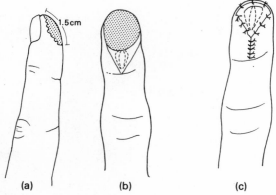

(a)   (b)   (c)

Fig. 16.
(a) Pulp amputation; oblique and more than 1.5 cm in length
(b) The V-incision, the broken line represents the deep neuro-vascular pedicle
(c) The V-shaped advancement flap, with pedicle, sutured in position with interrupted stitches, and Y-closure

possible, even at the cost of small additional shortening, because in the result they are superior to terminal scars or Wolfe-grafts. Re-implantation of a finger-tip works well in children under 2, but is time-wasting and disappointing in adults. Cf. **Wrist and hand amputations**.

Avulsion of fingers

This injury occurs in lathe and drill workers whose gloves, catching in the rotating component, are twisted rapidly with the finger inside them. The avulsion is usually at the metacarpophalangeal joint and the tendons and nerves may be pulled out at considerable length. Adequate closure of the skin and of the palm may require excision of the metacarpal head in order to avoid deformity and loss of the palmar cup.

Amputations

When you have decided that amputation is inescapable you have to choose between length and function of the resulting digits. Your choice will be guided by common sense and experience, but a few guidelines may help.

Thumb Maintain its length if possible, even if it involves plastic procedures such as transferred pedicle grafting to achieve skin cover. A full-length thumb with two stiff joints in the correct position, fully opposed to the index, may be more useful than a short thumb which moves a bit. It can be shortened later anyhow if necessary. Cf. **Wrist and hand amputations**.

Other digits If you amputate the middle and/or ring fingers at the metacarpophalangeal joints the patient loses his palmar cup and cannot hold his loose change or a handful of corn. Preserving even half the proximal phalanx avoids this.

If metacarpophalangeal amputation of one finger is inevitable, removal of two-thirds of the metacarpal as well will close the gap at the expense of the palmar breadth. Women usually prefer this as it gives a less unsightly hand. Manual workers need palmar breadth. Every case must be fully assessed on its merits and full discussion with patient and/or parents is very important. Specific written permission should always be insisted on before operation.

In general, amputation of the whole of the smallest digit

Collateral ligaments

in the hand are injured in side to side wrenching accidents. Those of the interphalangeal joints of the fingers should generally be repaired. Otherwise a troublesome instability remains.

The ulnar collateral ligament of the first metacarpo-phalangeal joint should always be repaired. Rupture is readily identified by firm radial deviation of the thumb whenever local pain, swelling, and tenderness suggest local injury. If it is not repaired, permanent, disabling instability, with loss of pinch, ensues. It needs early identification and early repair. Always look for it when examining a 'sprained' thumb.

More surgeons at present recommend a shortened thumb with a good volar flap as superior to a grafted one. Every case is different and needs individual consideration.

should be taken back to the proximal one-third of the fifth metacarpal as it is unsightly and incovenient at the metacarpophalangeal joint level.

The most careful consideration of every aspect of the patient's interest and welfare must always be taken before embarking on such a disfiguring operation. Anxious people often prefer to give an irretrievably injured digit 'a chance' before amputation, and such a request must be respected. If possible inevitable amputation should be carried out after 2 or 3 weeks and not longer.

High pressure injection injuries

The usual sequence of events is first the puncture wound, then an increasingly painful finger or hand, then progressive circulatory failure. The grease-gun is a common criminal and exerts pressures up to 70 000 kNm^{-2} (10 000 lb in^{-2}), producing immediate tissue necrosis and vascular coagulation. In addition grease may enter a tissue plane or tendon sheath and travel a long way.

Such destructive injuries need wide exploration and opening to the extreme limits so as to allow profuse irrigation with cetrimide solution (to which chlorhexidine 0.5% may be added).

Delayed amputation may be required but it should not be delayed too long. As soon as the destruction of the skin circulation of (for instance) a digit is established, amputation should be carried out to shorten the time of disability and allow an early return to work.

Another possible presentation is from a tractor-driver trying to staunch a leak from a hydraulic fluid line (e.g. operating a hydraulic lift) by digital pressure.

Oil-injection injuries

Common in intensive rearing systems for poultry where thousands of fowl-pest injections are given in a day's work. This inoculum is in an oily vehicle which causes local tissue necrosis. When an injection goes into a digital pulp, amputation may be required if the area is not treated as described in the preceding section. The antigenic component is relatively harmless to humans and causes no trouble, but the recipients show an encouraging rise in antibodies to Newcastle disease and its congeners.

All grey fat or ragged connective tissue should be ruthlessly de-brided until normal tissues only are left. Then the skin is loosely closed.

Abduction injuries of the deep tranverse ligament of the palm (intermetacarpal ligament)

A very painful injury which occurs particularly in Rugby players. Complete relief is given by Bedford elastic supports and healing is generally complete in 3 weeks. Residual discomfort may last much longer.

Blackthorn injuries

Common and very painful in hedgers. They always need careful exploration as the tiniest residue of blackthorn bark generally produces a painful aseptic abscess. The thorn is so tough and the hedger's blows so energetic that joints and articular cartilage are not uncommonly transfixed. The commonest places for the tip of the thorn to lodge are in a tendon (especially the extensor tendon over the proximal interphalangeal joint), a collateral ligament, or a joint capsule. Removal produces rapid relief of pain and swelling. Use a loupe.

HAY-FEVER

Can be of such acute and severe onset as to constitute a genuine emergency. When there is a combination of conjunctival chemosis with upper respiratory congestion, and sometimes wheezing as well, it is necessary to give urgent relief.

This situation arises commonly at the height of the grass-pollen season when contact has been close and intense (e.g. a sensitive child playing in long grass) and needs IV antihistamine, e.g. chlorpheniramine (Piriton 10 mg), followed by steroid injection (IM Depo-Medrone 40—80 mg) if relief is inadequate. If severe wheezing happens too, adrenaline may give the best relief (IV adrenaline 1/10 000, 1 ml given slowly).

Acute chemosis of the conjunctiva (so much more distressing to the observers than the patient) can be relieved by instillation of 1% adrenaline eye-drops (Eppy eye-drops), one or two drops at a time, until a response is achieved.

All such cases should be referred to their own physician for consideration for a desensitization course.

Optic chromoglycate recommended for chronic/recurrent sensitivity reactions of the conjunctiva.

HEAD INJURIES

The examination and assessment of head injuries in general and severe ones in particular are difficult and unreliable. Various writers have claimed to be able to make accurate prognoses on the basis of minutely collected physical signs, but experience suggests that these may be belied in the event. However, accurate recording of a full CNS examination on reception of patients with head injuries is important, and gives a useful base line for subsequent observations.

The key to this situation is to be found in the recorded changes taking place in routine observations accurately kept in the accident department and then in the intensive care unit during the first few hours and days after injury. Special attention has to be paid to changes in

1   Pupillary state and reaction (size in mm always)
2   Level of consciousness and response to stimuli
3   Respiration
4   Body temperature
5   Muscle tone and posture
6   Pulse and blood pressure

Your own hospital is certain to have appropriate charts for recording these. Progressive changes occurring in the first few hours can give an urgent indication for specialist intervention, possibly to relieve increasing intracranial pressure by means of burr-holes; but this is unusual. Early deterioration in (e.g.) deceleration injuries is generally due to diffuse cerebral changes that are not amenable to surgery. This, however, is not the place to give detailed advice, and reference should be made to standard works on head injuries and intensive care (see **Bibliography**).

The most crucial contribution to the recovery of severe head injuries that can be made in an accident unit is the immediate establishment of an airway and adequate oxygenation in all cases. This may involve cuffed endotracheal intubation on reception of the injured person, even laryngotomy (q.v.) in cases with severe mandibular or sub-mandibular injuries who have laryngeal occlusion (may be followed later by formal tracheostomy), and the provision of proper means of IV infusion.

Computerized axial tomographic scanning in early stages is of value but is not yet generally available.

Ultrasonic scanning of midline displacement can give early warning of an extradural haematoma if CAT scanning is not available. Cf. **Neurological emergencies**.

Cardiac or respiratory arrest have to be dealt with as they arise and are of very grave significance.

There is a good deal of dispute about the use of IV steroids where brain damage is severe, but there is evidence that adequate dosage (use e.g. 100 mg of dexamethasone or 500 mg of methyl-prednisolone IV) may reduce reactive oedema and thus improve cerebral blood-supply, and no clear evidence of any adverse effects. Hypertonic infusions* should only be given on the direct instruction of a neurosurgeon. Their effectiveness is not in dispute but is of a 'once only' kind and cannot be repeated.

It is worth saying again that the vital treatment of patients with severe head injuries consists in early intubation, careful airway maintenance, and adequate oxygenation. All the other things are extras. Cf. **Antibiotics.**

Minor head injuries

In a sense these are more of a problem than severe head injuries. Which do you admit? The safest rule is to admit all in whom there is significant alteration of consciousness and all in whom there is radiological evidence of recent fracture. Every minor head injury not admitted should leave with written instructions properly explained. A suggested form is as follows:

Minor head injuries
Rest for 48 hours after injury unless told not to
Report to the family doctor or to this department in case of:
  Abnormal drowsiness
  Persistent vomiting
  Disturbance of behaviour
Persistent headache after 48 hours unrelieved by aspirin should
  be reported to the family doctor

There seems to be no reason to admit the fractured skulls which present several days after injury with a sub-aponeurotic haematoma.

Head and multiple injuries

The prognosis of patients combining severe intra-cranial damage with other major injuries (especially intra-thoracic or intra-abdominal) is bad. Every case needs intensive care and the most painstaking assessment by all the members of the trauma team in closest collaboration.

*e.g. colloids such as Dextran 70 or 150, or Mannitol 25%.

In the absence of witnesses *amnesia* is the significant indication of concussion; post-traumatic amnesia is less significant than pre-traumatic.

HIP SYNDROME, PAINFUL

This condition, common in children, can sometimes be regarded as a simple traumatic synovitis. It is characterized by a painful limp. and relieved by not bearing weight. The leading physical sign is limitation of abduction and external rotation of the femur. It is often clearly unimportant, but in cases occurring between the ages of 4 and 9 years it should be regarded as a possible presentation of Perthe's disease. Occurring at puberty it may be due to slipped upper femoral epiphysis. In the first age group the hips should be X-rayed 1 month after the onset, if symptoms persist (with genital screening). Exposure to X-rays is contra-indicated at the onset. In the second age group early X-ray is important. Reference to an orthopaedic specialist is necessary if there are any suggestive bone changes in either hip. Treatment consists in encouraging rest and non-weightbearing until the pain and limp have subsided. If there is fever don't forget that osteomyelitis or septic arthritis are still possibilities. A simple toxic arthritis (e.g. concurrent with acute tonsillitis) explains some cases.

HYPERTHERMIA AND HEAT-STROKE

Due to derangement of thermo-regulation caused by salt and water depletion; may be made worse by infective pyrexia at the same time. Commoner in children in hot weather and still more so in those with fibrocystic disease. Rectal temperature 43–45 °C (108–112 °F). Rapid cooling by use of sheets soaked in cold water and dried by an electric fan should be continued till the rectal temperature falls to 39 °C. IV saline. Admit to ICU. If the temperature remains above 43 °C for long, irreversible brain-damage ensues. Too rapid cooling can lead to peripheral shut-down and consequently to a sudden rise of core-temperature. Tepid bathing and continuous rectal temperature-recording is probably the most effective regime. (*Cont. opposite*)

HYPOTHERMIA

is common after drowning in winter waters, lying drunk in a ditch all Saturday night, or solitary falling with an apoplexy or a fractured femur and not being found till next day by the neighbours. Its severity is estimated by taking a rectal temperature with a

(*Cont.*)

First-aid measures include oxygen by face-mask, IV hydro-cortisone 100 mg and, for rigors, IV calcium gluconate 10%: 5–10 ml or by infusion in glucose-saline.

A similiar condition (**malignant hyperpyrexia**) can follow anaesthesia with Halothane and other anaesthetic agents. This is associated with an autosomal dominant predisposition. Admit to ITU.

special low-reading thermometer, and any fall below 35 °C should be regarded as a demand for treatment. When the body temperature is below 30 °C the routine treatment by enclosure in a foil envelope may need to by supplemented by saline infusion heated in the warming circuit to 39 °C. When shivering begins you know that you are making progress, and the return of consciousness is often heralded by extreme restlessness, which can be confidently controlled by a small dose of diazepam (5–10 mg IM). Admission under the care of an appropriate specialist is always required. An electrical continuous reading thermometer with a thermal probe in the rectum makes management of such patients much easier.

IMPETIGO (cf. Skin emergencies)

Very common in children, often unrecognized, generally inadequately treated. Requirements are detergence with 1% cetrimide solution or cream, and antibiotic cream (neomycin/gramicidin combined – e.g. Graneodin) applied locally, both every 2 or 3 hours for 48 hours. If the condition is severe with signs of toxaemia, flucloxacillin by mouth (staphylococcal or streptococcal infection).

INFECTIONS, MISCELLANEOUS

Cellulitis, facial

Spreading from a pararhinal boil or angular stomatitis can rapidly produce cavernous sinus thrombosis and death if untreated. Formerly common but now only occurs from neglect or extreme poverty. Antibiotics, given so as to procure maximum blood-levels, cure it. Rupert Brooke, the poet, died of it.

Cellulitis, orbital

Secondary to antral or other sinus infection; poses a serious threat to the eye and brain. Treat as above and refer to ophthalmologist.

Dermatitis gangrenosa

Starts like a cellulitis (e.g. of the volar aspect of the forearm) and proceeds on its indolent and inexorable way to skin-death. Antibiotics are ineffective and skin-grafting is usually needed. Probably

On reception take venous blood for biochemical profile and full blood-count. Self-warming, minimum interference, and early admission to ITU are the key approaches.

Complications are ventricular fibrillation, acute pulmonary oedema, hypotensive shock, heart-block, sinus bradycardia, and acute pancreatitis. These are dealt with on their merits. Too rapid rewarming predisposes to cardiac arrhythmia.

immunological in origin with secondary infection. Combined systemic steroids and antibiotics worth a trial.

Erysipelas

Acute Streptococcal usually (Lancefield group A); intradermal infection with a palpable edge; formerly deadly but no longer so. Amoxycillin orally suffices to control it and local ichthammol applications, as compress or impregnated bandage, give relief of local burning pain.

Chronic and recurrent Usually in elderly faces and due to non-haemolytic streptococcus; responds well to co-trimoxazole.

Erysipeloid

Not very common; occurs in butchers and fishmongers, typically on dorsa of hands or feet. Indolent, itchy, and sore to touch. Responds to amoxycillin. Causative organism *Erysipelothrix insidiosa*.

INFESTATION (see Table 4, pp. 95–7)

Hardly accidental or very urgent in Britain but a common incident which has to be coped with in people attending A & E units for various reasons.

In temperate climates fleas, lice, and mites are the front-runners, ticks coming a poor fourth. Fortunately the first three are susceptible to γ-benzene hexachloride 1% in an appropriate form. If resistant strains develop, organophosphorus compounds like malathion 0.5% are effective but more toxic. Repeated applications may be required to fit the life-cycle of the parasite. Ticks are usually solitary or nearly so and dealt with mechanically (see **Sheep-ticks**). Benzyl benzoate for scabies is too toxic to skin for general use.

In tropical and sub-tropical regions the variety of infestations is much larger and includes numerous grubs, worms, tape-worms, protozoa, and so on. It is easiest to use a tabular presentation of different types, emphasizing aspects of the clinical manifestations of the parasites which are likely to be an occasion of emergency presentation. Two diagnostic features common to the large majority are fever and eosinophilia.

Cf. Fournier's gangrene: Slater and others, *Journal of the Royal Society of Medicine* **75** 530 (1982).

Liver abscess

Usually anaerobic infection with non-specific symptoms but generally showing enlarged liver on palpation and characteristic changes in hepatic enzyme-levels. Ultrasound scan can localize the abscess to allow confirmatory aspiration and identification of the pathogen. Aspiration and antibiotic treatment may eliminate the need for surgery.

An excellent review of 16 cases by Perera and others (*Lancet* **ii** 629 (1980)) gives a valuable understanding of the clinical variety and therapeutic approach in this life-threatening condition.

95

TABLE 4 RELEVANT PARASITIC INFESTATIONS

Common name	Causative organism	Regional prevalence	Clinical presentation relevant to A & E	Emergency treatment	Specific treatment	Disposal	Comments
Worms (helminths)	Nematodes						All worms Hygienic precautions of crucial importance in prevention; efficacy of treatment should be checked by appropriate pathological monitoring (e.g. ova in faeces)
Round-worms	*Ascaris lumbricoides*	Universal	Intestinal obstruction; volvulus	Admission to surgical care	Piperazine; Mebendazole		
Whip-worms	*Trichuris trichiura*	Tropical and sub-tropical	Dysentric symptoms in debilitated children	Diagnostic only	Mebendazole	Medical supervision	
Hook-worms	(a) *Ancylostoma duodenale* and (b) *Necator americanus*	Tropical and sub-tropical	Dysenteric symptoms in patients at risk; rarely intestinal obstruction after treatment	Rarely needed	(a) Bephenium hydroxy-naphthoate; (b) thiabendazole	Medical supervision	
Larva migrans	*Strongyloides stercoralis* (cf. *An-*	Tropical and sub-tropical	Acute pulmonary episodes (pneumo-	Rarely needed	Thiabendazole	Medical supervision	

					supervision	
pin-worms – vermicularis						
Elephantiasis *Filaria bancrofti (Wuchereria bancrofti)*	Tropical, world-wide	Acute lymphangitis or orchitis; secondary effects of chronic lymph-oedema	Diagnosis; general surgical principles	Diethylcarbamazine followed by IV suramin	Medical or surgical supervision	Spread by anopheles and culex mosquitoes
Brugia malayi; Brugia timori	Tropical Asia	Paroxysmal nocturnal cough, sometimes with haemoptysis	Admission	Diethylcarbamazine	Medical supervision	Spread by mansonia mosquitoes
Calabar swellings *Loa loa*	West and Central African rain forest	Subconjunctival worms (*microfilariae*)	Admission to medical care	Diethylcarbamazine	Ophthalmological supervision	Cf. onchocerciasis (*Onchocerca volvulus* – mainly blindness)*
Guinea-worm *Dracunculus medinensis*	Tropical, world-wide	Painful abscesses containing larvae or worms; monarthrosis of knee	Medical – 1 week before surgical – as needed	Diethylcarbamazine followed by traditional extraction by winding round a stick	Medical supervision	Water purification eliminates risk

*Vector – female black fly (*Simulium*); treatment Diethylcarbamazine-C and Suramin (new drugs expected)

96

TABLE 4 RELEVANT PARASITIC INFESTATIONS *(Cont.)*

Common name	Causative organism	Regional prevalence	Clinical presentation relevant to A & E	Emergency treatment	Specific treatment	Disposal	Comments
Tape-worms	*Taenia solium* (measly pork)	Universal in pigs	Cysticercosis (larval stage) in muscle, brain, meninges, liver, lungs; abscess-type of presentation; dead cysts calcify early	Complicated clinical features demand experienced specialist care	Niclosamide for intestinal phases	Special supervision	Rare except in people who eat underdone pig meat; very serious implications
Dog tape-worm	*Echinococcus granulosus*	Universal in cattle and sheep (larval stage); transmitted by sheep- and herd-dogs	Hydatid disease (larval stage) in liver (80%) and brain; ruptured hepatic hydatid cyst → acute peritonitis with severe immunogenic shock	Specialized surgical care only	Niclosamide if diagnosis of primary infection is made by screening the family of a known case	Special supervision	Australia, New Zealand, Wales and Scotland; excellent review article by Morris, *British Journal of Hospital Medicine* **25** 586 (1981)
Flukes	*Schistosoma*	Sub-tropical and tropical			Metiphenate; Praziquantel		

Bilharzia *haematobium*	Africa, many	Urinary symptoms, especially painful haematuria			supervision	Water-purification, persecution of snail-hosts, and public health education are specific
mansoni		Intestinal bleeding and ulceration	Surgery rarely needed	Niridazole	Medical supervision	
japonicum		Cerebrovascular occlusion		Organic antimonials	Medical supervision	
Liver-fluke of cattle *Fasciola hepatica*	Universal in herbivores on marshy ground	Hepatic symptoms (pain, tenderness, and cholestasis)	Medical admission	Niridazole	Medical supervision	
Protozoa *Kala-azar* *Leishmania donovani* (cf. **Skin emergencies**)	Tropical, world-wide	Chronic wasting fever with splenomegaly and late skin-lesions	Admission to medical care	Organic antimonial compounds; splenectomy	Medical supervision	Resistant Sudanese types treated with diamidine compounds
Malaria (malignant tertian) *Plasmodium falciparum*	Tropical and sub-tropical, world-wide	High pyrexia, collapse, delirium, and coma; splenomegaly (cf. **Haematological emergencies, Malaria**)	Diagnosis by thick blood-film; IV saline-infusion; admission	IV quinine 10 mg/kg followed by oral mefloquine; or 200 mg Nivaquin IV very slowly in 20 ml water	Medical supervision; suppressive routine	Splenic rupture commoner in more chronic forms of malaria

97

TABLE 4 RELEVANT PARASITIC INFESTATIONS (Cont.)

Common name	Causative organism	Regional prevalence	Clinical presentation relevant to A & E	Emergency treatment	Specific treatment	Disposal	Comments
Sleeping sickness	Trypanosoma brucei (tsetse fly)	African savannah	Coma	Supportive treatment	No specific treatment		
Chaga's disease	Trypanosoma cruzi (cat- and dog-bugs)	Central and South America	Lymphadenopathy and hepatosplenomegaly; cardiomyopathy	Supportive treatment	No specific treatment		Control of insect vector decisive
Toxoplasmosis	Toxoplasma gondii (cf. Respiratory emergencies)	Universal cat-transmission	Unlikely to present as emergency*	None	Spiramycin and systemic steroids; pyrimethamine and triple sulphonamide	Medical supervision	
Visceral larva migrans	Toxocara canis	Universal cat- and dog-transmission; children especially	Abdominal pain; choroidoretinitis; asthma;	None	Carbamazepine	Medical supervision	Animal hygiene, especially domestic; child hygiene

dysentery	histolytica	where water supply and sewage are confused	colitis can → gangrene of colon; amoebic abscess of liver	warm-stool microscopy; admission; IV saline	and tetra-cycline	supervision to ensure clearance
Miscellaneous *Fleas, lice, and scabies*						See preamble to **Infestation** above
Jiggers, or chigoe fleas	*Tunga penetrans*	Dry, sandy soil in tropics, world-wide	Itching ulcerated and secondarily infected soles; tetanus a common complication	Local antisepsis, systemic antibiosis, and healing care *secundum artem*	No specific treatment	May need admission in severe sepsis; Traditional folk-medicine often successful
Tumbu bites	*Cordylobia anthropophaga*	Tropical Africa	Larval abscesses similar to warble in cattle	Gentle surgery, antisepsis, and antibiosis for secondary infections	None	Simple supervision till healed

Advice about *tropical infestations* is available in the UK from: Liverpool School of Tropical Medicine (051) 708 9393; London Hospital for Tropical Disease (01) 387 4411; East Birmingham Hospital (021) 772 4311.

*But incidental discovery of choroido-retinitis or cerebral calcification may call for comment; may present with pyrexia and malaise only.

INJECTIONS

It is curious how much confusion of thought there is about methods, sites, and effects of different methods of injection. There are 4 types of injection in daily use:

Intradermal

For epidermal anaesthesia in opening superficial abscesses, or for test-doses of possible allergens.

Subcutaneous

For producing slow onset, long-acting effect — e.g. antigenic inoculation or long-acting analgesia.

Intramuscular

For medium onset, medium duration, and relative freedom from pain when injecting irritant solutions such as penicillin. The three preferred sites are:

> The deltoid muscle (*not* distal to the muscle where the radial nerve lies!)
>
> The upper and outer quadrant of the buttock (*not* in any of the other quadrants, which contain fat rather than muscle, not to mention the sciatic nerve)
>
> The antero-lateral aspect of the thigh at half its length — i.e. lateral to the subsartorial canal, medial to the fascia lata, and into the vastus lateralis muscle

The buttock is the best site for the injection of antibiotics.

IM injections in shocked patients may of course not be appreciably absorbed owing to peripheral circulatory shutdown. When normal circulation is restored they will then exert a delayed, unwanted, and probably unrecognized effect.

Intravenous

For rapid onset, short duration. This is the best method for relief of pain in accidental injury. It gives a quick relief and in cases of abdominal injury makes proper examination possible (without it this may not be possible at all) and in doubtful cases is of short

duration so that minor or developing abdominal symptoms and signs are not masked for long.

INJURIES, MINOR

Nail in board penetrating sole

Very common indeed. Every case should have IM penicillin and cloxacillin or other appropriate antibiotic, local tincture of iodine, and be off the foot for 3 days. These measures are justified by the frequency of septic complications and high rate of disability. Tetanus toxoid antigen as indicated.

Fork through foot

Treat as above, for the same reasons. Exploration of severely contaminated wounds is needed.

Ruptured plantaris tendon (so called)

See **Emergencies, miscellaneous**.

Ruptured biceps (long head)

Spontaneous, in elderly men as a rule. Symptomatic treatment.

Bites of tongue

Very common in children. Do not suture unless there is uncontrollable bleeding or gross dehiscence. Sutures in the mouth are uncomfortable and inefficient, and generally get infected. If you have to stitch, use 4/0 or 5/0 Dexon, not cat-gut.

See also **Dental, miscellaneous**.

INSANE, CERTIFICATION OF THE

This tends to be increasingly difficult and complicated and there is some doubt whether house officers are entitled to use Sections of the Mental Health Act. In cases of extreme urgency where

Trichanchonē ('hair-strangling')

Also known as *blanket gangrene*; not uncommon in babies who wriggle their toes. Strands of wool, cotton, or synthetic fibres encircle the toes (fingers rarely), get pulled tight, produce oedema and circulatory occlusion; eventually gangrene. Treatment is only effective if a longitudinal linear incision down to bone is made in the affected digit. Relief is rapid. Choose a plane without crucial structures in it. Antibiotics etc. as common sense suggests.

Antibiotics for 48 hours and frequent mouth-washes.

Cf. **Psychiatric emergencies** and **Confusion, acute**.

there is danger from a violent psychotic, ring the police and request them to remove the patient to a place of safety under Section 136 (some feel unable to do this as they regard the A & E unit as a place of safety). From there the Social Services Department should cope. In cases where the patient is merely in need of care and treatment for mental illness, *either* contact the GP and ask him to invoke Section 29 of the act, *or* ask the senior casualty officer (if he is approved under Section 28) to cope, *or*, failing both these:

1 Ring the ambulance station for the name and number of the social worker on duty
2 Contact the social worker
3 Contact the psychiatric consultant about admission to the nearest mental hospital
4 Wait and hope

The ramifications are endless, but these two sections are likely to be the relevant ones. Avoid sedation if you can, so that the executives can draw their own conclusions from an unfuddled patient.

INTUSSUSCEPTION

Should be considered in every vomiting child under five in the presence of colic and the absence of diarrhoea. Red-currant jelly on the tip of the examining finger is a positive confirmation but its absence is not significant. Cf. **Abdominal injuries**.

KNEE

Acute effusion

May be due to traumatic synovitis *or* to haemorrhage. Apply pressure bandage and review next day, when the distinction will be clear. Aspirate the really tense ones via the supra-patellar pouch — lateral approach. Simple effusions are cool; haemorrhagic ones are hot. Pyogenic arthritis needs emergency admission (cf. **Musculo-skeletal emergencies**, p. 106).

Internal derangement of the knee

Especially cartilage injuries. Localize the injury by examination.

Admit for observation or urgent surgery.

Sir Robert Jones's original bandage has been modified to: first layer of Tubigrip (or equivalent) not too tight; second layer of plentiful cotton or wood-wool; third layer of Tubigrip over all.

Give IM pethidine 100 mg, and see if the rest will straighten the knee. If genuine locking occurs (which is fairly rare), manipulate under IV narcanalgesic cocktail (see **Narcanalgesia**) or general anaesthetic. POP cylinder or pressure bandage gives support and confidence after reduction. If either of the collateral ligaments or the cruciates have gone, a pressure bandage will suffice till the morning unless the knee is unstable, when admission is necessary. POP cylinder with weightbearing for 10 days to 3 weeks is the definitive treatment for injury to either collateral ligament and should not be delayed longer than is essential.

Different orthopaedic specialists have different ideas and you must always try to treat in the way they prefer. Some prefer to deal with knee disorders from the beginning and may wish to perform examination under anaesthetic and/or arthroscopy. Remember that arthroscopy may be more painful and slower to recover than the original injury.

Repair of collateral ligaments which are ruptured (estimated by instability of the joint on lateral stress) is needed.

Haemarthrosis

can occur without apparent structural damage to the knee, especially in the elderly and arthritic. If it occurs assume that there is structural damage until the contrary is proved. If it is tense, aspirate with full aseptic precautions.

Lumps in popliteal fossa

Semimembranosus bursa   Medial.

Popliteus bursa   Lateral.

Baker's cyst   Degenerative, with arthritis; usually midline.

Popliteal aneurysm   Pulsatile.

LARYNGOTOMY (crico-thyroidotomy)

is an invaluable emergency operation which is very rarely necessary except in cases of mandibular fracture associated with injury to the floor of the mouth. Still more rarely it may have to be resorted

In order to exclude the presence of a loose body you need antero-posterior and lateral radiographs with obliques, tunnel, and sky-line views as well. These may also disclose *osteochondritis dissecans* of the medial femoral condyle (e.g.).

Pain on lateral stress is usually absent if rupture is complete.

If blood is withdrawn leave it to stand; a supernatant layer of fat globules is very suggestive of bony injury even if X-ray is negative. POP cylinder and rational follow-up is specific.

to in cases of asphyxia where intubation proves impossible. It is done as follows:

1 Palpate the cricothyroid membrane with one finger: it is above the ring of the cricoid and below the inferior notch of the thyroid cartilage.

2 Make a *transverse* skin-incision over it, 2.5 cm wide.

3 Clean the cricothyroid membrane by blunt dissection with a dry gauze swab.

4 Press the sharp end of a number 15 scalpel blade through the membrane so as to make a *vertical* slit.

5 Insert a medium-sized disposable tracheostomy tube (Portex 27 French gauge) through the slit and tie tapes round the neck — or a child's inflatable endotracheal tube which, with any luck, will allow positive-pressure ventilation.

6 Secure haemostasis by pressure or rarely by use of clamps — this is the last item; oxygen, suction, anaethesia can follow through the airway thus provided as the occasion demands.

Provided laryngotomy is not maintained for more than 48 hours, and provided the crico-thyroid membrane is opened with a *vertical* incision, subsequent difficulty with phonation is said to be entirely avoided. Every casualty officer should be expert at this operation and able to perform it in under 2 minutes, if required. (It is generally easy to arrage to practise on cadavers. The operative incision is simply concealed by a small piece of strapping and no disfigurement is caused.)

LUMBAGO

is a name given to the following clinical syndrome: characteristically the complainant is youngish, the onset of lumbar pain is sudden and unrelated to injury, and there is spasm of the lumbar muscles, which typically begins with a trivial action (e.g. stooping to pick something up). Cases of lumbago often present themselves as 'back injuries' and are all too often accepted as such. The history is crucial.

The pathogenesis of such lesions is unknown and will probably remain so, because they are not fatal. It is sometimes said that they are due to interfacetal connective tissue disorders in the lumbar spine. This cannot be disproved.

Various laryngotomy tubes are coming on the market. Choose the
best.

The treatment is simple where muscle spasm is severe: IV muscle-relaxant, e.g. methocarbamol 1.0 g (Robaxin 10 ml) given slowly over 3 or 4 minutes. Once this has been done it is possible to assess straight-leg raising reliably, and so differentiate from a true prolapsed intervertebral disc, and secondly to perform such manipulations as may seem desirable to overcome or alleviate the underlying condition. Subsequent treatment should consist of: combined relaxant and analgesic by mouth, e.g. Robaxisal-forte (methocarbamol with aspirin); reasonably brief rest, using floor or fracture boards if pain is severe; physiotherapy, especially dorso-lumbar extension exercises — these are essential, as good muscle tone is the best (or only) prophylactic against recurrence. If the condition is associated with well marked local tenderness in (e.g.) an interspinous ligament or muscular insertion, generous infiltration of the tender area with local lignocaine 1% and/or Marcaine 0.5% (bupivacaine) can give dramatic relief. Do this considerately so as not to hurt your patient.

If the syndrome does not approximate to the above description, other possible sources of pain should be investigated and eliminated. The muscle-relaxant treatment gives excellent and rapid relief if used appropriately. It is useless for miscellaneous chronic low backaches. If it is not successful, further radiological, haematological, and physical examination is required.

If investigation leaves the patient with pain but without specific diagnosis, *manipulation* (q.v.) under anaesthesia is often effective.

LYMPHANGITIS

Superficial

Common and familiar concomitant of boils, whitlows, etc. Treat the cause. Rest the limb.

Thrombosing lymphangitis

Less common: infection of main lymph vessel either proximal to infection (as above) or in a retrograde direction from infection, e.g. axillary abscess — characteristically there is a tender string stretched down the medial side of the arm towards the elbow, which markedly curtails arm movements, especially extension and

abduction. The condition responds to IM flucloxacillin, and if untreated leaves considerable disability, sometimes for several weeks.

MAJOR CIVIL ACCIDENTS (in USA 'mass casualties')

May descend upon a hospital with only an hour or two's notice. Aircraft and train accidents can bring overwhelming numbers of casualties, and so can football crowds and terrorist attacks by fire or explosives.

There is no way of making adequate preparation and the more complicated a provisional plan is the greater the confusion that will ensue.

The essential features of any plan are:

1 The recipient of the alert (the telephonist?) should have a short list of key people to tell.
2 The key people need a short list of the second echelon to tell. All these people need to know beforehand in writing what their duties are.
3 Ward-space has to be cleared.

Apart from communications, the aim of forward planning should be to prevent chaos by appointing one person to receive and document casualties as they arrive (usually the A & E Consultant or his deputy), an experienced surgeon to establish priorities for treatment, and a reliable physician to deal with medical emergencies. The next aim is to allot one junior doctor and one nurse to each major casualty for resuscitation and supervision and hope that there are enough surgeons available to do the necessary operating.

Anything beyond this is trimmings.

MALARIA

Plasmodium falciparum infection may present as an acute pyrexial emergency with collapse. History may be confused by the prolongation of incubation period (normally 3 weeks) owing to drug prophylaxis. Emergency treatment consists in prompt IV saline infusion and admission for continued supportive care and specific

Malaria prophylaxis

Prophylaxis may be asked for in accident units. The current recommendation is proguanil 100 mg daily or chloroquine 500 mg weekly in areas of chloroquine sensitivity, Maloprim (pyrimethamine 25 mg + dapsone 100 mg), one tablet twice a week, in areas of chloroquine resistance. This is not good for pregnant women, who should, if possible, keep out of malarial areas under the influence of chloroquine-resistant plasmodia. (See Mackay, *Prescribers' Journal* **20** (6) 137 (1980).)

Quinine 10 mg/kg IV in 250 ml saline is advised for cerebral malaria (malignant tertian ague) followed by Fansidar or tetracycline to prevent recurrence.

chemotherapy. Thick and thin blood-films should be made on reception, to identify the parasite. Cf. Table 4, p. 96a.

MANIPULATION

is of two kinds:

Diagnostic

This is well exemplified by the painful knee in which specific diagnosis may be impossible because of pain and muscle spasm. General anaesthesia gives relaxation so that *gentle* manipulation can identify crepitus, locking, clicking, instability, and abnormal movement. Then a reasonable attempt at diagnosis can be made.

Therapeutic

Applicable to a variety of A & E presentations, especially necks, backs, elbows, knees, and feet, not to mention a wide variety of dislocations (q.v.).

Once a careful history and relevant investigations have established the absence of significant musculo-skeletal disorder, or mere stiffness, as a source of pain and disability, whether due to trauma, disuse, misuse, or some other obscure (possibly psychogenic) cause, *manipulation* by experienced hands can give rapid and effective relief. It is an art more than a science and better taught by apprenticeship than talking or writing, but has to be controlled by reason and appropriate scepticism.

I use it for occipital neuralgia, acute torticollis, low back-ache, painful shoulders, tennis elbow, locked knees, stiff feet, and painful toes.

Success depends on knowing what you are at, being clear about your aim, and understanding from experience how much force to use.

If you manipulate a severely spondylitic spine, acute supraspinatus syndrome, vertebral Pott's disease, unstable dislocation, or some other risky pathological state, you will richly deserve the heavy damages that will be given against you in court.

The essence of a therapeutic manipulation is the enforcement of maximal movement of the relevant joints so as to relieve their stiffness and increase their mobility.

Benign tertian (*P. vivax, P. ovale*) and quartan (*P. malariae*) malaria need chloroquine to treat the acute attack. *P. vivax* hypnozoites in the liver (the origin of relapses) are killed by primaquine (Bell, *Prescriber's Journal* **23** (5) 122 (1984)).

Recent work from China suggests that Qing Laosu may offer fresh hope for cerebral malarial patients afflicted with chloroquine-resistant strains (*Fifty-Two Prescriptions* (168 BC)). It is derived from *Artemisia annua* (Bruce-Chwatt, *BMJ* **284** 367 (1982)). Mefloquine is a new synthetic from the US Army, not yet established.

P. vivax can incubate for 9 months and relapse for 9 years. If blood-films are negative but clinical suspicion is strong, bone-marrow smears may give positive findings (Simmons and others, *BMJ* **284** 113 (1982)).

Relaxation of muscle spasm shortly ensues.

MUSCULO-SKELETAL EMERGENCIES

Mostly dealt with under separate headings (see **Backs, Crystal Synovitis, Dislocations, Elbow, Emergencies, miscellaneous, Hand infections and injuries, Hip syndrome, painful, Knee, Lumbago, Manipulation, Necks, Neuropathies, Soft tissue injuries, Tendon injuries, Tenosynovitis, Torticollis**; for individual conditions see index).

Thumb, ulnar collateral ligament rupture

Occasionally sustained in awkward falls, holding reins or ski-sticks especially. The metacarpo-phalangeal joint is painful and swollen and completely unstable on radial deviation. The patient cannot pinch effectively. Primary repair is called for and needs an experienced operator; if not repaired the instability and disability are permanent. It is a condition frequently missed and should be looked for specifically in every case of sprained thumb.

Painful shoulder

Shoulder pain is a rotten pain, and the end result of severe shoulder pain is all too often a 'frozen' one. Careful differential diagnosis and early treatment should go far to prevent this.

Arthritis, osteo-   Keep it moving by support, encouragement, physiotherapy and anti-inflammatory drugs.

Arthritis, pyogenic   Needs drainage, antibiotics and punctilious after-care. Even so, since it occurs usually in the elderly and debilitated, a stiff joint is often the outcome.

Rotator cuff syndrome may follow a fall, a minor strain or even come out of the blue. It is an acute synovitis/capsulitis involving the tendons of the rotator cuff which surround the joint and constitute the greater part of its structure. Rest, anti-inflammatory drugs and mobilization from the third and fourth day may produce mobility. If not, early intra-articular steroid injection should be given – either into the subacromial bursa or the sub-deltoid (see Fig. 17). Direct injection into the front of the joint tends to be too painful.

Supraspinatus syndrome is a localized variety of the above and

Frozen shoulder is an adhesive capsulitis characterized by severe pain on attempted external rotation of the humerus. Its course is benign but prolonged (3–12 months); intra-articular steroids are generally useless; oral naproxen 250 mg in the morning and 500 mg each evening helps a bit. Maintaining morale is a problem in severe cases. Not usually traumatic in origin.

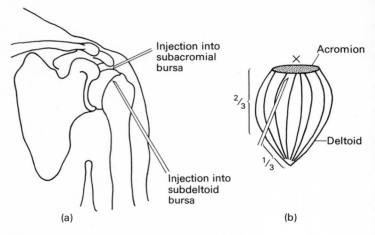

Fig. 17.
    (a) Two approaches for intra-articular injection of steroids; the sub-deltoid approach needs a spinal needle
    (b) Scheme of deltoid muscle — point of entry of needle, aimed at the centre of the acromio-clavicular joint

merits a similar approach. Characterized by the 'painful arc' on abduction, or, if severe, by loss of abduction. If there is accompanying calcification of the tendon a 10-day course of oral steroids in diminishing dosage (starting with, e.g., prednisone 5 mg three times a day) should follow the intra-articular injection of Depo-Medrone (2 ml).

Stable, impacted fracture of the humeral neck is best treated as a soft tissue injury, especially in the aged. Rest for three or four days and then mobilize gently but rapidly. Broad arm sling helps to avoid disimpaction.

Painful foot

Stress fracture (q.v.)

Morton's metatarsalgia   Interdigital 'neuroma'; activated by bilateral compression by the metatarsal heads. May be relieved by topical steroids and rest; may need surgery. Orthopaedic problem.

Plantar fasciitis   Sometimes generalised, more often locally tender under the heel. Foam padding in *both* shoes, anti-inflammatory drugs or topical steroid injection in sequence or together.

Hallux rigidus   A degenerative consequence of a congenital abnormality of the 1st metatarso-phalangeal joint. Exacerbation often follows trauma. Limitation of flexion more than extension. Flexion deformity occurs late. Recognition and palliative treatment allay anxiety, but orthopaedic surgery is often needed in the end.

Osteochondritis

Instances of this curious group of developmental disorders of bone often present primarily in the A & E department. They are probably determined by local varieties of vasculitis in growing bone and are mostly self-limiting. A few need treatment.

A list is a useful aide-memoire, together with their eponyms as these are still generally in use.

> *Carpal lunate*, Kienböck's: may lead to necrosis and the need for excision of the lunate. Recognized by increased radiological density.
>
> *Hip*, Perthe's disease of: see **Hip syndrome, painful.**
>
> *Knee*, osteochondritis of femoral condyle: may present as a loose body after 'dissection' of the affected area of cartilage.
>
> *Patella*, chondro-malacia of: characteristically tender on grinding. Rest in POP cylinder. Surgery rarely needed.
>
> *Tibial tuberosity*, Osgood and Schlatter's: POP cylinder.*
>
> *Tarsal scaphoid*, Köhler's: symptomatic treatment only.
>
> *Heel*, Sever's: symptomatic treatment only by raising the heels.
>
> *Metatarsal head*, Freiberg's: often needs excision.
>
> *Vertebral bodies*, Scheuermann's.

Osteochondritis dissecans can affect the elbow (capitellum), ankle (talus), or hip. Medial femoral condyle commonest site.

These conditions form a strange collection and may not all be causally related.

MYOCARDIAL INFARCTION (cf. **Cardiac arrest**)

1   Make a diagnosis. This is often difficult and not necessarily

Sinding—Larsen—Johansson — distal pole of patella: symptomatic only.

*Symptomatic only.

clinched by electrocardiogram in the acute phase. A clinical
hunch may be all you have: never ignore it.

2 Give sedation and/or analgesia: pentazocine (Fortral) 60 mg,
pethidine 100 mg, or cyclimorph 15 mg (all the benefits of
morphia without its central emetic effect). Diamorphine should
not be used in an accident department in my opinion.* IV
diazepam (at least 10 mg), if more sedation is needed.

3 Send for a physician. If there is marked dysrhythmia start an
IV infusion with normal saline and run it in slowly. This is a help
for IV medication, but it is important not to produce a circulatory
overload. Alternatively use one of the numerous IV cannulae with
a side-arm for intermittent medication. If you have connected the
monitor to the patient, the arrival of numerous infra-nodal
extrasystoles (see Fig. 18) is an indication for IV lignocaine at
once, either direct (lignocaine 1% 2 ml IV) or in the drip (lignocaine
5% 10 ml in 500 ml of infusion fluid). Bicarbonate infusion is
used after arrest is reversed to correct the ensuing acidosis. A
cautious IV bolus of 50 ml of sodium bicarbonate 8.4% is safer in
the emergency unit than continuous infusion. It is easy to correct
harmful acidosis to equally harmful alkalosis. The bolus should
be given by way of the IV cannula and washed through by the
infusion fluid. IV infusion is not a routine requirement; reassurance, calmness, quiet, and sedation are. Atropine is effective for
sinus bradycardia; isoprenaline for heart-block. In an emergency
either can be given IV in a dosage judged by its effect. Arrhythmias
and supraventricular tachycardias are more safely dealt with by IV

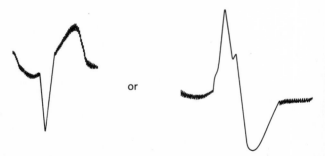

or

Fig. 18.

*Because it can attract heroin addicts posing as victims of infarction, and because it gives some people the horrors.

Pseudo-coronary: spontaneous rupture of oesophagus while vomiting against a vigorously contracting crico-pharyngeal sphincter; perforation is usually 2 cm above diaphragm into pleural cavity; if higher, can cause mediastinitis. Early sign is emphysema of neck. Urgent radiological diagnosis and urgent surgery. First aid is urgent treatment for *toxic shock* (q.v.) with IV Haemaccel, steroids, and IV antibiotic. Dopamine as indicated.

The logic and effectiveness of intracardiac calcium chloride (13.4% − 10 ml) in asystole, followed by intra-cardiac adrenaline (0.1% − 1 mg, i.e. 10 ml), is in doubt. An alternative approach is to give intracardiac atropine (0.6 mg) followed by a 400 J d.c. shock − this can be repeated once. Bicarbonate as above. Pacing via an oesophageal pacing-catheter may be a new way forward. Survival rate at present is around 1 per cent.

lignocaine 50 mg in a bolus dose than by the bewildering multitude of autonomic moderators which can, in inexpert hands, themselves cause arrhythmias and asystole. Disopyramide may be an exception, but fashions in this field change so rapidly that it is wiser to keep it and the current β-blockers and unblockers in the poisons cupboard for the use of visiting physicians.

NAILS, DIGITAL

Injuries

Crushing and laceration involving finger-nails in particular can be misleading, as a simple cut or split in the nail may conceal important underlying injury to the nail-bed or terminal phalanx. If there is any doubt the whole nail should be carefully dissected off the subjacent tissues and the damage properly inspected, assessed, and treated. Careful suture may be required, or even reconstruction, in order to avoid prolonged disability or later complications.

Ingrowing toe-nail

is a tiresome, common, and disabling complaint attributable partly to tight footwear (socks as well as shoes), partly to bad nail-care, partly to congenital conformations of the toe conducive to this disorder.

Recurrent affliction should be dealt with by preventative packing of the corners of the growing nail so as to lift it over the paronychial pads, careful trimming of the nail so that it ends in a strictly horizontal edge, and, if nothing else will serve, extirpation of the germinal layer by careful carbolization of the ungual matrix. This must be done using phenol 90% for 3 minutes and close attention to detail is essential. The method is safe, effective, relatively painless, and quick to heal. It needs to be done directly after removing the nail and in a bloodless field. This procedure entirely replaces the tiresome and inefficient methods of surgical extirpation of the matrix. Meticulous care is essential for success.

Unilateral ingrowth is dealt with by unilateral resection of one quarter of the nail and carbolization. Bilateral ingrowth by bilateral resection of one quarter. Total ablation remains as the only cure for cylindronychia, pachyonychia, and onychogryphosis (Fig. 18a).

Pacing can be life-saving in extreme bradycardia of sudden onset. In the emergency unit trans-oesophageal pacing is quickest and safest (e.g. Oesocath-Vygon).

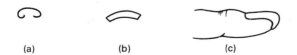

Fig. 18a. (a) Cylindronychia. (b) Pachyonychia. (c) Onychogryphosis. (a) and (b) transverse schema of nail-deformity.

Subungual haematoma   Paracentesis unguis (with white-hot wire; electric cautery, not a paper-clip and spirit lamp), preferably in cases where there is no associated fracture. Subsequent dressing required to mop up slow leakage of the haematoma, which often persists for 48 hours. Relief is usually immediate. If it remains painful after paracentesis, look for a fracture of the terminal phalanx.

NARCANALGESIA

When urgent manipulations are needed, e.g. in dislocations which threaten the circulation in a limb, or when an anaesthetic is not available to help in dealing with other painful injuries, this method is invaluable in procuring completely reliable amnesia and reasonable muscular relaxation. It is safe if administered with adequately painstaking care (see **Diazepam**).

Narcanalgesic cocktail (for an adult)

  Cyclimorph 15 mg

  Midazolam (Hypnovel) 2 ml: dose can be repeated × 1. Give 0.5−1 ml at a time with 5 minutes between. It is a good muscle-relaxant. Midazolam has a shorter half-life than diazepam and consequently a much shorter post-narcotic confusion. It is rather slower to give, but for out-patient use distinctly safer. It is similarly reversible by doxapram. It should replace diazepam in this procedure.

  Naloxone 0.4 mg

  Doxapram 200 mg post-operatively: each bolus is given IV and chased by a few mls of normal saline.

This method must be administered with every precaution (see **Diazepam**). Its main advantages are the lower dose of diazepam needed and the consequently shorter recovery time. The naloxone acts slowly in reversing the narcotic and analgesic effects of morphia, and protects the respiratory centre from depression; it should be given immediately the painful stimulus has stopped.

NECKS

Acute neck pain, often with signs of root irritation, is a common emergency. Cf. **Torticollis**.

Cervical prolapsed intervertebral disc

This may follow an accident, sudden movement or strain, or nothing out of the way. Root pain is the characteristic, generally referred to the suprascapular area; sometimes down the arm.

See **Neuropathies**. Severe cases may need admission for continous traction.

X-ray is essential and the illness may last for several weeks.

Occipital neuralgia

is a common consequence of whiplash injuries and blows to the head. Onset is usually delayed for a day to two and the chief symptom is persistent frontal headache unrelieved by analgesics. Tenderness on local pressure over the great occipital nerve is diagnostic.

Manipulation under general anaesthetic or IV narcosis, after X-ray, gives early relief. Forced rotation is the key manipulation.

Cervical spondylosis

can accompany any of the acutely painful necks and sets a problem to the manipulator. If it is severe enough to threaten the stability of the spine, firm support is the only treatment. Experience and orthopaedic consultation must decide.

NEUROLOGICAL EMERGENCIES

Rarely a therapeutic commitment to A & E units – but it's as well to identify the varieties and treat the treatable with understanding.

Cerebro-vascular accidents

may require airway care or intubation. Oxygen should always be given, as for head injuries.

Cerebral haemorrhage/thrombosis/embolism

Subarachnoid bleeding   Neck rigidity and blood in cerebro-

The basic therapeutic regime is hard collar (adjustable type) by day; also by night if tolerated; if not, use soft collar (firmly applied) at night. When the acute symptoms and muscle spasm subside start traction or manipulation as circumstances suggest. Collars work by splintage and, if not applied so as to immobilize the cervical spine, do no good.

Ultra-sound or heat may be helpful.

When permanent support is needed, an individually moulded head and neck support is best (e.g. Orthoplast or Plastazote).

spinal fluid. Find out local policy about IV cyclokapron as prophylaxis in this disaster.

Sub- or extra-dural bleeding   Traumatic: sub- — over 45; extra- — under 45.

Extra-dural haematoma, from leakage from the middle meningeal vessels, is identified by local swelling, radiological fracture, dilatation of pupil on the same side, increasing headache and irritability giving way to coma. Lucid interval usual. Prognosis fairly good. Absolute emergency.

Sub-dural haematoma, from lacerations of the venous sinuses, generally starts soon after injury. There may be a lucid interval or not. Progressive signs of rising intracranial pressure with less marked lateralization. Generally there is considerable brain damage. Prognosis poor. Relative emergency.

Pyogenic conditions

Cerebral thrombo-phlebitis   E.g. cavernous sinus thrombosis. Associated with general septicaemia or pyogenic conditions (cf. **Infections, miscellaneous**, Cellulitis, facial).

Brain abscess   May mimic acute confusional state or neoplasm.

Neoplasm   Haemorrhagic or oedematous crises.*

Primary infections

Pyogenic meningitis
Viral meningitis
Early *poliomyelitis* in invasive stage
Encephalitis — bizarre visual phantasies
Tabetic crises — severe fleeting pains in limbs or abdomen; Argyll—Robinson pupils and syphilitic neuropathy

Miscellaneous

Herxheimer reaction from treatment of active syphilis
Myasthenic crisis
Guillan—Barré syndrome — immunologically determined poly-neuritis going on to
Bulbar palsy

Diagnostic aids are: any lateralizing signs; ultra-sound scanning of the mid-line echo; best of all, CAT scan of head. Expert neuro-surgical advice needed. Cf. **Head injuries.**

*Preceding history of cerebral disturbance may be absent.

Myopathy is not primarily an A & E concern but can present with acute respiratory crises due to choking, respiratory insufficiency, or injury. Treatment is common-sense application of routine measures.

Acute vertigo — viral (labyrinthitis)/acoustic neuroma/idio-pathic: if you treat this with prochlorperazine (Stemetil) do not forget that this can produce acute extra-pyramidal symptoms, usually facial athetosis, reversible by benztropine (Cogentin) and withdrawal of the prochlorperazine.

NEUROPATHIES

Various kinds may present in A & E units from long-tract lesions (syringo-myelia and multiple sclerosis) to motor neurone disease and amyotrophic lateral sclerosis, but their emergencies are incidental rather than accidental.

Prolapsed intervertebral disc (cf. Backs, Necks)

The commonest neuropathies seen are those arising from spondy-lotic changes and intervertebral disc-lesions. Cervical and lumbar disc calamities often present as 'injuries' but a careful history discloses a practically spontaneous onset in a majority.

Acute cervical disc-lesions with root-signs (pain) but no muscle weakness are often effectively relieved by the following regime:

Adjustable collar by day

Soft collar by night

(both must be properly applied)

Analgesics ⎫ for 1 week
Rest   ⎭

followed if needed by

Cervical traction

Continued support

Gradual weaning from the use of collars is important.

Lumbar disc-lesions can be identified by limitation of straight-leg-raising and their progress monitored by observing its recovery. Rest in bed for 3 weeks on fracture-boards is good treatment but limiting. POP jackets may take 10 days to give relief but are very good in the long run and permit busy people to keep working. Temporary lumbo-sacral supports (e.g. Remploy) can be very useful in less severe or recurrent cases.

Not every case needs full orthopaedic investigation and sup-port, and those that do may well find it difficult to obtain.

The same applies to the femoral stretch-test.

Rarely central disc-prolapse can cause acute retention of urine.

Orthopaedic consultation may however be necessary and must not be forgotten.

Careful examination, X-ray and regard to the therapeutic outcome offer early relief to people suffering from a wretchedly painful and disabling complaint. These cases are well within the capacity of any painstaking GP but unfortunately the therapeutic facilities seldom are — one reason why the national average for annual increase of attendances at accident units in Britain stands at 10 per cent! Cf. **Lumbago**.

Saturday-night palsy

Radial palsy due to compression of the radial nerve in the musculo-spiral groove of the humerus, historically attributed to falling into a drunken stupor with the gloriously relaxed arm flopping over the arm of a chair. It can also occur at the axillary level in the same way (cf. 'crutch palsy').

It is treated by the use of a Brian Thomas splint (Fig. 19) and may take weeks or months to recover.

Bell's palsy (cf. **Viruses**)

Painless and sudden onset.

Zoster of the geniculate ganglion (Ramsay–Hunt syndrome)

Look for the tympanic vesicles which soon merge into a general tympanitis. There is generally pain before palsy (cf. **Viruses**). Adreno-corticotrophic hormone 40 international units daily or at 48-hour intervals.

Acute eighth-nerve palsy (viral)

See **Ear, nose, and throat emergencies**.

Thoracic outlet syndrome

Nerve-root compression by cervical rib, fibrous bands, clavicle on first rib (shopping-bag palsy); also common in pregnancy. Sling gives first-aid relief. It is characterized by aching pain (worse by day) in the early stages, and interosseous wasting later.

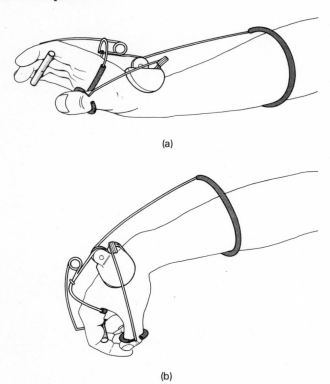

(a)

(b)

Fig. 19. Brian Thomas splint. (a) Position of relaxation. (b) Position of use.

Other tunnel syndromes

Elbow Anterior and posterior interosseous nerves; ulnar nerve.

Carpal tunnel syndrome Median nerve pain in palm, spreading up the forearm; worse at night. Relieved by night-splintage. May need surgery.

Popliteal tunnel syndrome Intermittent claudication in young soldiers.

Anterior and posterior tibial compartment syndromes Rare and needing surgery.

Meralgia paraesthetica   Lateral cutaneous nerve of thigh compressed at the pelvic brim or in the muscle belly of sartorius. May need a surgical remedy.

Peroneal palsy   Surgical remedy.

NOMENCLATURE

It is common for the obsessional to be meticulous about labelling, and this is not necessarily useful. But in an accident department, where cases are continually passed from one casualty officer to another, accurate labelling is important and time-saving. This is particularly true when medical reports have to be made in the absence of the patient and after the relevant doctor has left. Incidentally, if your labelling is accurate your diagnosis is likely to be the same. When in doubt please refer to *Gray's anatomy*.

Fingers

As there is no consistency in the numbering of fingers it is easiest to use translations of the old anatomical labels (from radial to ulnar):

Thumb
Index
Middle
Ring
Little

It has been known for judges in court to get very savage with opposing medical witnesses who refer to the 'second' and 'third' fingers, when they both intend to refer to the identical digit.

NURSING SERVICES

The success of the A & E enterprise is dependent on the skill and quality as much of its nursing as of its medical staff. The relationship between them is important. In most departments (unless they are attached to a teaching hospital which supplies students in quantity) experienced nurses need to take a large share of daily routine work. As well as dressings and general organization they often undertake suturing, venepuncture, cardio-

graphy, catheterization, monitoring, plastering, and even intubation. Very well they do them too. It has to be remembered that
procedures outside their general training remain the responsibility
of the medically qualified staff, but nevertheless they are done to
a standard which often exceeds anything that medical staff are
capable of. A female nurse tackles the stitching of wounds often
far more skilfully than a male doctor, and her patience and skill
in dealing with multiple lacerations experienced in RTA is well
up to the standards of the best plastic surgery. Nurses' training
in these extra tasks is the responsibility of the senior casualty
staff and no pains are too great to be expended on it. They will,
of course, gain greatly from the nurses' skills, but the chief
beneficiaries will be the injured. Distinction of function between
nursing and medical staffs (but *not* distinction of responsibility)
rapidly becomes blurred in accident work, and the more blurred,
the better the department works. It is important for all parties
in the team to work actively towards a relationship of constructive
harmony.

OBSTETRIC EMERGENCIES

In countries where hospitals are far apart and medical services
thinly spread any obstetric emergency can turn up. Ideally an
obstetric surgeon should always be available to deal with impacted
labours, maternal or foetal distress, major obstetric haemorrhages,
mal-presentations, and foetal deformities. In fact accident
surgeons will have been conditioned by experience to cope when
they are unsupported.

The major obstetric emergency in Western Europe, where (in
general) populations are indulged with specialist support for
every crisis that misfortune can bring, is the death of a mother
in the last trimester of pregnancy following (typically) a RTA.
Any experienced accident surgeon or casualty officer should be
prepared to do an instant Caesarian section to extract a live baby.
Thereafter the problem is one of ordinary infant-resuscitation,
warming, and disposal to an appropriate baby-unit. The practical
problems and pitiful heart-break which ensue are not the casualty
officer's problem.

Accidental haemorrhages (concealed or revealed) and ante-

A current epidemic of nursing disorder characterized by preoccupation with politics, doctrinaire policies irrelevant to patient care, and manipulation of clinical situations for power-oriented motives is sweeping through UK hospitals. It is wearisome, harmful, and destructive. It is essential to believe that it is self-limiting and that the centuries-old tradition of selfless patient-service will re-establish itself, or be re-established.

partum haemorrhage may also arrive at our door and may need IV analgesic/sedation and the institution of resuscitation by IV infusion before being sent to a specialist unit. The casualty officer's response will be determined by the time-lag involved in the transfer. If there is far to go or long delay in getting there, then resuscitation and cross-matching at least may be needed.

Cf. **Abortion.**

OESOPHAGEAL OCCLUSION

is a common A & E calamity, because it is of sudden onset (though possibly preceded by dysphagia, q.v.), often brings great discomfort, and is always frightening. **Diagnosis**: retrosternal pain may be low (achalasia or neoplasm) or middle (mediastinal new-growth) or high (oesophageal pouches and diverticula); indirect laryngoscopy shows salivary pooling in the piriform fossae; water is regurgitated. The neoplastic and neuromuscular causes of this condition are obvious, but the 'greedy old man' syndrome is less well known, though much commoner. It consists in swallowing unchewed lumps of meat (usually) which the elderly oesophagus cannot transmit. Generally they soften physiologically and pass on their way. Sometimes they have to be extricated by way of an oesophagoscope. Some swear by the efficacy of passing a naso-gastric tube or a small balloon-catheter which can pass the obstruction and then be withdrawn after inflating the balloon; this then acts as an extractor. Such remedies may well be of use in the absence of a capable gastroscopist but could well do irrevocable damage to an oesophagus obstructed by stricture or neoplasm. The prevention of recurrence is not by admonition but by mincing all meat. Pouches and diverticula can occlude the gullet by simply filling slowly with food, drink, and spittle until nothing more can get by. Regurgitation may relieve the blockage, but oesophagoscopy and repair may be required.

ORF

A virus infection of the facial skin of sheep, occasionally transmitted to shepherds and children. Its appearance is characteristically that of a small, *painless* carbuncle, of slow onset. Secondary

Precipitate delivery and post-partum haemorrhage may come urgently to A & E. Syntometrine (ergometrine maleate 500 μg + oxytocin 5U) is the best emergency drug and should be given IV after the delivery of the anterior shoulder. It should also be given IV for post-partum bleeding without attempting to identify or remove retained placenta.

Eclampsia Fulminating hypertension arising out of 'toxaemia of pregnancy'. Heavy sedation (IV midazolam 5−10 mg); darkness and quiet; urgent specialist management. Presenting feature: fits.

Incarcerated retrogravid uterus, usually with acute retention of urine. Catheterize; possible replace in anteversion if specialized help is far away.

Prolapsed cord Sedation and Trendelenberg if available. Otherwise elevate the pelvis on pillows. Urgent referral to specialist unit.

infection is generally a presenting symptom, possibly with lymphangitis and lymphadenopathy. Treatment is by IM antibiotics for the secondary infection and idoxuridine in dimethylsulphoxide for the primary lesion. It is slow to develop and fairly slow to go. Most commonly it occurs in the hand. Idoxuridine is specific for the primary lesion; 15% is generally sufficient and is applied for 10 minutes once or twice daily for 3 days as a wet compress.

Differential diagnosis

Carbuncle   Always painful.

Vaccinia or paravaccinia   History and virological studies. The viruses of orf and paravaccinia are identifiable (for practical purposes) by electron-microscopy.

Cellulitis or other epidermal lesions. By profile, which is characteristically raised in orf (see Fig. 20(a)) as opposed to the profile shown in Fig. 20(b), which characterizes the ordinary boil.

PAIN

used to be the physician's first enemy, but death has displaced it. It is still important to postpone death when possible, but it is sometimes forgotten that pain should be got rid of as the first step of all. Every patient brought in by ambulance or car in great pain should be given brief examination and relief *before* he is lifted out, put on a trolley, and stripped for examination. In general IV pentazocine (Fortral) 15–60 mg is sufficient – IV

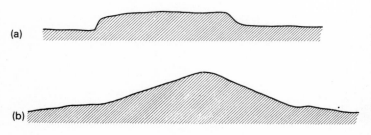

Fig. 20.

Erythema multiforme is a common complication which merits a short course of oral steroids. It can be very severe and distressing.

because analgesia is quick and disposal rapid. The injection will often make proper examination of the injured and frightened possible when without it it is impossible. Pethidine (25–50 mg) is better for the aged as it will not make them vomit. The best general narcotic in my experience is Cyclimorph (morphia 15 mg with cyclizine 50 mg), which can be followed by naloxone 0.4 mg if the respiratory centre is at risk. Some surgeons are annoyed when patients are given this necessary relief, but the damage is to their *amour propre*, and not to the certain diagnosis of the patients' disease or injury.

Entonox (50–50 mix of O_2 and N_2O) should be available in every A & E department and can be used freely without risk. Many ambulances carry it and use it effectively for patients in pain during their journey. It is good for painful dressings after surgery or injury.

PARAPHIMOSIS

must be reduced even if there is not urethral occlusion, as swelling can occur rapidly. The standard textbook reduction by two thumbs is often effective after IV narcanalgesia. If this is unsuccessful a dorsal slit should be performed with blunt-ended scissors after infiltration with lignocaine.

In children general anaesthetic is needed. The dorsal slit operation is safer and more effective and less liable to recurrence than the standard operation of dividing the compressing ring. Subsequent elective circumcision may be needed to tidy up unsightly flaps.

PHYSIOTHERAPY

This is a hospital department which has its own expertise and pride and should never be used as a dump for patients in whom you have failed to make a diagnosis or cure.

Physiotherapists have special tools (e.g. ultra-sound), special skills (e.g. manipulations), and inexhaustible patience. They can and do help greatly with sprained ankles, acute spinal disorders without injury, stable fractures of tarsal and carpal bones, which are painful, and rehabilitation of patients whose injuries have

been treated by immobilization. They also have an invaluable contribution to make in helping the frightened and hurt to use joints which have become stiff and painful for a variety of reasons, and only need movement to make them comfortable and effective once again. They are an integral part of the accident family and appreciate being consulted, shown films of injuries, and being asked for advice. Their role can be improved and enhanced by their being included in relevant clinical trials (q.v.).

'PLASMA EXPANDERS'

In cases of blood loss you need often to use a 'plasma expander' to fill the gap until blood becomes available. While you are wondering which to use put up a litre of normal saline, which cannot do any harm unless the patient is in congestive cardiac failure.

There are three choices: Human Plasma Protein Fraction as supplied by the Blood Transfusion Service (this has now replaced dried plasma and carries a lower risk of plasma-transmitted hepatitis), which is costly and not really appropriate for simple bloodloss; a low-molecular-weight polysaccharide (Dextran 40); or a gelatin solution such as Haemaccel.

Human Plasma Protein Fraction should be kept for patients with plasma loss (i.e. extensively burnt patients), and the Dextrans have the disadvantage of interfering with blood cross-matching, preventing normal clotting and bleeding mechanisms from working effectively, and, by producing extra-cellular fluid retention, causing pulmonary oedema.

The gelatin solutions have none of these snags, have a short half-life (4 hours), are diuretic, and produce no ill effects on normal clotting and bleeding mechanisms. They are made from degraded beef-gelatin (which produces a short-chain molecule) and are not antigenic. The anaphylactoid reactions to these solutions occasionally reported in the medical press are said to be due to increased histamine production in the stomach, not to immunological explosions. They can be treated by IV cimetidine 200 mg (10 mg/kg). In use they have proved effective and trouble-free.

Haemaccel is therefore the 'plasma expander' of choice. Inci-

dentally, it is not very expensive when supplied on contract, has a shelf-life of 8 years, and is unaffected by freezing or tropical temperatures. Another make is Gelofusine.

Do not forget that blood-loss of more than 1—2 litres needs urgent replacement by blood. Nothing else will do.

PNEUMOTHORAX

Traumatic
See **Chest injuries**.

Spontaneous

In the young Common after exertion; characterized by sudden pain in the chest radiating towards the back. There may be considerable cardio-respiratory distress at onset owing to mediastinal mobility. X-ray. Refer to physician for admission if extensive (more than half of the hemithorax or whatever the locally accepted ruling may be). Spontaneous resolution is usual and thoracic intubation is rarely needed.

In the elderly Ruptured emphysematous bullae in cases of cor pulmonale may tip the balance of cardiac sufficiency quite drastically, and produce an acute cardio-respiratory emergency. Oxygen and paracentesis pleurae for first aid. Then admission: to intensive care unit if respiratory embarrassment is severe.

POISONING

Common in children by accident and in adults on purpose. In 1970 it accounted for 50 per cent of acute female admissions at Addenbrooke's Hospital, Cambridge. The numbers still do not seem to fall. The **management** is basically simple as far as primary treatment goes. It consists essentially of

1 Re-establishment of vital functions (intubation, IV infusion, cardiac support)
2 Removal of any poison that may still be in the stomach (stomach wash out or emesis) — saline emetics never to be used as they are themselves eletrolytically toxic and an

occasional fatal result is reported, especially in children; syrup of ipecacuanha is the best emetic for general use.

3 Administration of a specific antidote if available
4 Elimination of what has been absorbed
5 Adsorption of poisons in the bowel where possible
6 General supportive care

Exceptions to these general rules are:

Corrosives, which must not be treated by gastric lavage or emesis, as they may well intensify the damage done to the gullet on the way down, as they come up again, producing perforation where only ulceration had occurred before. Buffer solutions should be given either by mouth or by Ryle's tube. Milk is the best, but ice-cream is more acceptable to small children. If there is evidence of perforation already (great retro-sternal pain and collapse) nothing must be given by mouth but reference made for immediate oesophagoscopy, etc.

Volatile hydrocarbons such as petrol, benzene, carbon tetra-chloride, chloroform, and trichlorethylene should not be treated with whole milk as their absorption is increased by the presence of fat in the small bowel. This probably applies to kerosene (commercial paraffin) as well. These should be treated with copious watery fluids or skim-milk by mouth and *admission*, as toxic effects may be of late onset. The danger of inhalation of these agents during vomiting is of prime importance, so that stomach wash-out and emesis are contra-indicated.

Children, in whom some consider that gastric lavage should never be used because of the stress and anxiety it causes them. Emesis should be obtained by giving ipecacuanha. Others do not entirely accept this view, at least in the case of iron or aspirin overdose. Iron, they hold, is so toxic that no delay in emptying the stomach is permissible, and aspirin is notorious for forming a hard ball in the stomach, which can delay absorption by as much as 12—24 hours, with surprising and possibly fatal delayed intoxication. This is a trap for the unwary.

Obscure poisons

These often set a problem of a particular kind because the agent may be unfamiliar both to you and to the expert toxicologist —

even direct reference to a manufacturer may produce no information whatever — or even of unknown constitution. A few suggestions on how to proceed may be useful in the circumstances, and the following approach is suggested:

1 At first contact (usually by telephone) insist that the container, contents, and any available maker's instructions are brought with the patient: this is very important and can easily save hours of further messing about.

2 Known pharmaceutical preparations and galenicals: refer to standard works (*The British National Formulary*; *Treatment of common acute poisonings*).

3 Known agricultural poisons and sprays: identify in *Approved products for farmers and growers*.

4 If no definite information is available look up in *Clinical toxicology of commercial products*, identify the constituents of the poison, and refer to them in the appropriate sections.

5 Ordinary domestic preparations thought to be toxic: treat as 4.

6 If all else fails ring one of the poisons centres (p. 128) and if necessary winkle out the toxicologist. These centres have a definite but limited usefulness: they can give useful information in many cases, but cannot in general discuss problems of diagnosis and management. Their standards are high and information extensive. The responsibility for diagnosis and treatment remains, however, with the clinician.

7 Every accident department should keep a toxicology file to give access to up-to-date poisons information — recent information and practice is not necessarily the best, but it must be familiar and easily available: refer to it whenever in doubt.

Specific antidotes

Opiates: naloxone 0.4 mg IV repeated every ten minutes or so till desired effect is produced. Also effective against alcohol and a wide range of narcotics, including dextropropoxyphene.

Iron: desferrioxamine — procedure:

1 Take blood for serum-iron

Dextropropoxyphene causes respiratory depression. Naloxone is routine for all cases of Distalgesic poisoning.

2 Stomach wash-out or emesis to reduce scarring
3 Give desferrioxamine IM at once (0.5 g under 2 years old, 1.0 g over 2 years old); IV desferrioxamine 15 mg/kg is also recommended; probably best dealt with after admission, together with monitoring of serum free iron
4 Give desferrioxamine 5 g by naso-gastric tube
5 IV fluids if necessary: this can usefully convey desferrioxamine 15 mg per kg per hour by the IV route to a maximum of 80 mg per kg per day

Desferrioxamine (Desferal) is to be kept in a refrigerator.

Organophosphorus (e.g. Parathion, Malathion, Metasystox, sheep-dips, warble-washes and other insecticides): Pralidoxime (P_2S) — procedure:

1 Take 10 ml untreated blood for pseudo-cholinesterase-level in all suspected cases.
2 Inject pralidoxime 1.0 g IV and repeat if necessary.
3 Atropine 2 mg IM or IV for symptomatic control of anti-cholinesterase effects.

Heavy metals
  Lead: battery-burning
  Mercury: antifungal dressings
  Gold: therapeutic
  Arsenic (rat and weed killers): admirably dealt with in *Treatment of common acute poisonings* (see **Bibliography**, p. 167)
Dimercaprol should be available as an effective antidote even though these poisons are rarely met at present and are unlikely to present as emergencies.

Paracetamol is a dangerous poison which has been widely recommended as a safer substitute for Aspirin. It was introduced about the same time as Phenacetin was suppressed, in spite of the already established fact that their hepato-toxic effects are identical and severe.

It has been shown that amino-acids of the glutathion-precursor group effectively prevent hepato-toxicity if given within 10 hours of ingestion of the poison. Oral methionine is probably effective if given in doses of 2.5 g by mouth (or naso-gastric tube) 4-hourly up to a total of 10 g. IV N-acetylcysteine (150 mg/kg IV in 15

Organochlorines (e.g. dicophane (DDT), aldrin, dieldrin, gammexane, chlordane, heptachlor, endrin, endosulfan, etc. — persistent insecticides) Symptoms: mostly CNS with apprehension, excitement, and muscular fibrillation. Convulsions may ensue. Hyperpnoea leads to respiratory depression and apnoea.

Remedies Decontamination for accidental, and gastric lavage for intentional, poisoning; supportive treatment and sedation (e.g. diazepam) for convulsions. Toxicological consultation in obscure cases.

minutes, chased by IV dextrose 5%) can be used in cases of severe poisoning or those who vomit the methionine. See Fig. 21.

If treatment cannot be begun until after 10 hours, reference to a renal unit with a charcoal column to adsorb the poison from the blood should be urgently considered. Use a helicopter if necessary. Delay can prove fatal.

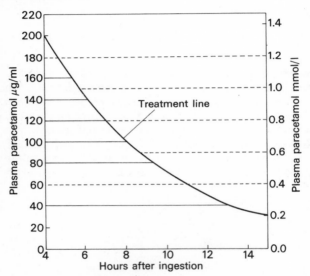

Fig. 21. Plasma paracetamol concentrations after overdosage. Treatment is indicated in patients with concentrations above the treatment line.

Phenothiazines (e.g. Maxolon, Stemetil): specific antidote is benztropin 1 mg/ml (Cogentin) for extra-pyramidal effects (see **Neurological emergencies**, Vertigo).

Tricyclic antidepressants: Merck, Sharp, and Dohme have recommended IV physostigmine salicylate 1–3 mg as a reverser of both cardiac and CNS effects.

Cyanide: amylnitrite inhalation as first aid; then give IV chelating agent Co-EDTA (cobalt edetate: Kelocyanor) 500 mg, followed by a further 250 mg 1 minute later if not effective. This is toxic

Oral charcoal must not be given with either remedy as they will be adsorbed by it.

Basic blood-paracetamol on reception the rule.

Note necessity of following nitrites with thiosulphate to correct methaemoglobinaemia.

and may induce collapse and vomiting for a brief period. Follow with 50 ml of 50% glucose IV. Secondly give amylnitrite inhalation and 25 ml of sodium nitrite 3% slowly IV. This is toxic too. Thirdly give 100 ml sodium thiosulphate 50% very slowly IV. So is this toxic. These quasi-specific antidotes should be prominently displayed in the reception area as speed of treatment is of paramount importance.

It is doubtful whether any other effective antidotes are at present available, but some may turn up.

Antidotes, general

Charcoal (activated and finely divided) can be given in water down a stomach or naso-gastric tube as an adsorbent and is reputedly effective in reducing the absorption of tricyclic antidepressants and paracetamol. Whole-blood filtration through a charcoal column is recommended for Paraquat and Diquat poisoning, and is available at New Cross Hospital (phone (01) 407 7600), East Birmingham Hospital (phone (021) 772 4311), and other centres which may be nearer to your own locality. It must be done as soon as the poison is identified in the urine, and if the patient is far away helicopter delivery is likely to be required if irreversible lung-damage is to be avoided. If you have seen the utterly hideous consequences of Paraquat poisoning you will readily agree to grasp at any straw which might deliver your patient from the delayed, horrific, and inevitable consequences of his folly or despair.

Fuller's earth and **Bentonite** are other adsorbent media which have been recommended for Paraquat. 150 g in milk, water, or squash can be given at first contact without fear of harm and with possible benefit. The toxicological hunt still has to go on just the same afterwards.

Poisons centres

New Cross Hospital Poisons' Unit, London: (01) 407 7600*
Cardiff: (0222) 569200
Edinburgh: (031) 229 2477 Ext. 2233
Belfast: (0232) 240503
Dublin: (0001) 723355

Mushrooms and toadstools

A review article ('Mushroom poisoning', *Lancet* **ii** 351 (1980))
gives an illuminating summary of the state of play. *Amanita
phalloides* is blamed for 90% of deaths, and one in three of
those poisoned by it will die. An educational effort to make
the olive-green cap, white gills, and basal 'veil' familiar to children and collectors is a simple and useful undertaking. Diagnosis
can be established early by history and detection of amatoxin
(not a generally available test) in blood, urine, and gastric juice.
Stomach wash out, charcoal by mouth, IV fluids, and electrolytes
with glucose, haemodialysis for the severely poisoned (more than
two stools eaten), and chemotherapy with penicillin (thioctic
acid (150 mg twice daily) and silymarin − neither available in
UK). Atropine (2 mg doses IV every 10 minutes till full atropin-
ization is reached − i.e. pupils are fully dilated and unreactive
and the mouth is dry) may give significant relief to the early
cramps and profuse diarrhoea and vomiting.

IV fluids to be started early to maintain urinary output and so
increase excretion of paraquat (William and others, *Lancet* **i** 627
(1984)).

A new gleam of hope from Cardiff for treatment of paraquat
poisoning: irradiation of affected lungs with low doses of γ-rays
labelled with cobalt-60. One successful case: *BMJ* **288** 1260
(1984).

Mannitol 20% is the vehicle currently recommended for Fuller's
earth; 250 ml given by naso-gastric tube (ICI 1980). Followed by
magnesium sulphate by mouth to increase excretion of adsorbed
paraquat.

*Or (01) 639 4380.

Leeds: (0532) 32799
Manchester: (061) 740 2254

Poisons bibliography
See general **Bibliography**.

PROCTALGIA FUGAX

Severe spasmodic anal pain generally at night. May be associated with constipation or overtiredness. Apply cold water to the anus, with instant relief. If it is not 'fugax', pain may be due to organic disease which needs surgical investigation.

PSYCHIATRIC EMERGENCIES

Acute confusion

whether emotional, toxic, or senile in origin, should be left untreated if possible until seen by a psychiatrist. If not it is best controlled by IM diazepam 10 mg, repeated SOS; or, if the restlessness is combined with violence or dangerous wandering, paraldehyde 10 ml from a glass syringe may be needed. Drug addiction is a major factor in big cities. Cf. **Drunkenness**.

Acute disturbance

Generally of emotional origin associated with crises of bereavement, matrimony, or other close personal relationships; can take many forms, viz. withdrawal, disorientation, hysterical fugue with loss of memory, etc. These must be handled gently with understanding from the very beginning. They do not need, as a rule, any emergency treatment, but gauche or insensitive handling in the A & E unit can increase the difficulties of all concerned to a serious degree. Drugs to be avoided in the A & E unit.

Acute mania

Usually associated with organic mental illness: manic-depressive psychosis, schizophrenia, cerebral tumour, etc. Also common in alcoholism (acute withdrawal mania, i.e. *delirium tremens*) and

Acute mania can be due to acute toxaemia (usually bacterial and post-operative) and to thyrotoxicosis (q.v.). Recognition early is crucial as the cause needs treating more urgently than the symptoms.

drug addiction either as part of a withdrawal syndrome or as a toxic manifestation of drugs like lysergic acid (LSD). Paraldehyde may be necessary to get the situation under control, but more importantly large doses of chlorpromazine (200 mg ½-hourly) are needed. The earlier IV high-potency B-vitamins are given in *delirium tremens* the better. (Cf. **Poisoning**, Specific antidotes, Phenothiazines, p. 127.)

Psychopathy

is a term applied to varieties of personality disorder which are a personal or social nuisance. Aggressive and manipulative psychopaths are the bane of accident units and are not usually violent. They require endless patience and skill to handle and are inaccessible to reasoning or pharmacy. The final resort is to ask for help from the police who often know all about them. They can disrupt an entire department for long periods, but seldom transgress the law.

Sociopathy

is an etymologically bastard form of the above in which the personality defect expresses itself in an inability to fit in with social norms or to make ordinary provision for basic needs such as food, shelter, clothing, and care of children. Such people haunt hospitals demanding fares to non-existent relatives in remote towns and hope to be able to spend the proceeds on (for instance) booze. Social workers should cope with them but in fact, more often than not, lack the facilities to do so.

PYREXIA OF UNKNOWN ORIGIN

Unravelling pyrexias of unknown origin (PUO) is no part of the duties of a casualty officer, but it is important to bear in mind the possibilities at least. Some pyrexias need rest in bed, aspirin, and fluids; some need urgent life-saving treatment; others need equally urgent isolation for the protection of contacts. It is not helpful to admit a case of influenza to a medical ward or a prodromal measles to a children's ward full of immuno-suppressed leukaemics. Likewise it is a pity to send home a *Plasmodium*

IV chlormethiazole (700–1500 ml) (Heminevrin) is the current remedy for *delirium tremens*.

falciparum malaria or an early case of rabies on symptomatic treatment.

Diagnosis, or at least reasonable suspicion, is an obligation in these circumstances. Particularly is this so in the case of travellers far from their usual source of medical care. They are the people who are likely to present with PUO and likely to be the very ones to be bringing home exotic and potentially dangerous infections from their travels.

Tabular presentation of different classes of pyrexia can act as a memory-jogger and arouser of clinical suspicion. See Table 5, pp. 132—4.

Do not forget that comparatively benign infections can prove disastrous to diabetics, immuno-suppressed patients, and the poverty-stricken or debilitated.

QUINSY (peritonsillar abscess)

A very painful and distressing emergency treated as follows:

1 Penicillin-G 600 mg IM at once or cephradine 500 mg
2 Surface anaesthesia to the pharynx with 4% lignocaine spray (X 3), the patient sitting up on bed or trolley
3 One stab in the area shown in *Pye's surgical handicraft*, using a scalpel with a number 15 blade on it: let the patient cough and spit
4 Open the incision widely with a small artery-forceps

Pus will pour out and the patient will be grateful for ever. Continue appropriate local bathing and systemic antibiotics. A good head-light is essential.

It may be that your ENT consultant will prefer to provide this treatment but maybe not.

RABIES

A canine zoonosis, enzootic all over Asia, America, and the continent of Europe. It is a neurotropic virus infection which is susceptible to treatment in the first few days but not afterwards. It can be transferred by biting, licking, or the mere contamination of skin wounds or mucous membrane with infected saliva. Incubation 10 days to 20 years; usually one or two months. The terror

(*Cont. on p. 135*)

132

TABLE 5   PYREXIA OF UNKNOWN ORIGIN

Class	Type	Clinical presentation	Management	Comment
Virus diseases without early localizing signs	Influenza A and B (with antigenic variations)	T 38–40 °C; aching head and limbs	Rest, fluids, and aspirin	Self-limiting
	Para-influenza 1, 2, and 3	T 36–38 °C; usually with laryngitis	Steam etc.; see Croup	Ditto
	Coxsackie A and B	pyrexia ± pharyngitis; intercostal myalgia	Symptomatic	Epidemic myalgia or Bornholm's disease due to Coxsakie B
	Epstein–Barr – glandular fever	Infinitely variable; usually pyrexial onset; confluent tonsillar exudate; palatal haemorrhages	Symptomatic; hepatic complications	Also called infectious mononucleosis; can be fatal
	Varicella – chicken-pox	Pyrexia; exanthem in 3–4 days	Symptomatic	Centrifugal rash
	Morbilli – measles	Pyrexia; exanthem in 3–4 days	Symptomatic	Koplik's spots precede rash
	Rubella – German measles	May present as polyarthritis about 10th day	Protection of contacts of child-bearing age	Occipital glands
	Variola major }smallpox Variola minor	Said to be extinct	May be fulminating	Centripetal rash
	Infective parotitis – mumps (myxovirus)	May not be pyrexial	Oral hygiene and rest	Complications common: orchitis, meningitis, and pancreatitis

Disease	Clinical features	Treatment	Notes
Ebola virus disease – Central Africa	intestinal upset; typical rash	Ditto	Haemorrhagic symptoms universal
Lassa fever – West Africa (Arenovirus)	Ditto; Fever may last 3 weeks	Ditto	
Hepatitis A	Fever; malaise; large tender liver; jaundice rare	Ditto	Gamma-globulin protects contacts
Hepatitis B – Australia antigen	Blood-handler's disease; infinitely variable course; immuno-suppressed patients	Vaccine available (1984)	High-titre immunoglobulin for patients at risk
Hepatitis non-A non-B	Blood-recipients and haemodialysis patients	Ditto	May lead to chronic liver disease
Arbovirus – yellow-fever and dengue	Yellow-fever mild to fulminant; central Africa and Central America; Dengue: haemorrhagic fever, Asia only; tick-borne encephalitis in European forests	General support	Mosquito-transmitted; no effective treatment; vaccination
Rabies (q.v.)		Hyperimmune globulin	Available from Immuno AG Vienna, (0732) 458101; Slow incubation 10 days to 20 years
Cytomegalovirus – common in immuno-suppressed patients	Glandular-fever-like illness	Symptomatic; anti-viral agents probably useless	Virus infection in very young, very old, debilitated, or immuno-suppressed patients can have dire consequences
Congo/Crimean haemorrhagic fever: see addendum, p. 134a			

133

TABLE 5 PYREXIA OF UNKNOWN ORIGIN *(Cont.)*

Class	Type	Clinical presentation	Management	Comment
Diseases due to intermediate organisms without early localizing signs	Chlamydia – psittacosis; cat-scratch fever; lympho-granuloma venereum	Insidious onset and prolonged course	Various combinations of tetracycline, erythromycin, and chloramphenicol	
	Coxiella – Q-fever; *C. Burnetti*	Ditto + cough		
	Mycoplasma – atypical pneumonia	Cough and malaise		
	Rickettsia prowazeki	Louse vector; blood culture	Disinfestation and relentless war on animal vectors; tetracycline and chloramphenicol combined	'Typhus'
	Rickettsia typhi	Rodent vector; milder form		
	Rickettsia rickettsii	Tick vector; Rocky Mountain fever		
	Rickettsia tsutsugamushii	Mite vector; scrub typhus		
Protozoal diseases with non-specific febrile onset, Nematodes, Filariae, etc.	See **Infestation, Malaria**			

febrile onset

Disease	Onset / incubation	Treatment	Notes
(...ore CHOLERA)			Still important in poor communities
Diphtheria	Short incubation	Tetracyclines and some other antibiotics; treatment requires specialist supervision	
Dysentery (bacillary)			Shigella and campylobacter dysenteries
Gonococcal infection in women (pelvic stage)			See Venereal disease
Légionnaire's disease (legionella pneumophila)		Erythromycin ± Rifampicin; early detection vital	
Leprosy	Slow onset		Special treatment and supervision
Leptospiroses L. icterohaemorrhagiae (Weil's disease) L. canicola L. hebdomadis	Short incubation	All treated with prolonged, high-dose penicillin	Rat-bite; contaminated bathing Dogs and pigs Field mice and voles; found in cattle handlers
Melioidosis (Pseudomonas pseudomallei)	Subclinical infection common; clinical presentation protean with predominant lung disease	Tetracycline and chloramphenicol	Asia and Pacific; comparable to the equine glanders
Plague (bubonic, pneumonic, and septicaemic)	Short incubation	Streptomycin and tetracycline; cf. Food poisoning, p. 59	Pasteurella pestis; vaccination (Culter Labs) (02814) 5151
Salmonella group	Intermediate incubation		
Tuberculosis	Slow onset	2- or 3-drug-combined treatment	Special treatment and supervision
Tularaemia	Slow progress from bite to lymphadenopathy; intermediate incubation	Biopsy and serology; European type-B minor; American type-A major	Treatment: streptomycin; tetracycline

134

TABLE 5 PYREXIA OF UNKNOWN ORIGIN *(Cont.)*

Systematic diseases due to varied organisms	Suggested areas of diagnosis	Comments
Respiratory	Atypical pneumonias; tuberculosis; lung abscess; bronchogenic carcinoma	Most of these need specialist diagnosis and treatment and are accordingly outside the scope of this book; see separate entries where relevant
Gastro-intestinal	Missed appendix abscess; ascending cholangitis; pyogenic or amoebic liver abscess	
Cardio-vascular	Subacute bacterial endocarditis, myocardial infarction	*Streptococcus viridans* subacute commonly; *staphylococcus aureus* usually acute and associated with valve prostheses
Uro-genital	Pyelonephritis, peri-nephric abscess	
Neurological	Viral meningitis; encephalitis	
Haematological	Septicaemia (especially meningococcal); sickle-cell crisis; malaria; pyaemia	
Dermatological	Erysipelas; impetigo; pemphigus; epidermolysis bullosa	
Connective tissue (usually immunological)	Rheumatic fever; Henoch–Shönlein purpura; polymyalgia rheumatica	
Oto-rhino-laryngological	Sinusitis, especially ethmoid	
Oral	Dental sepsis; Vincent's angina	
Immunological (usually affecting	Polyarteritis nodosa; disseminated lupus	

Drug reactions	E.g. sulphonamides
Blood transfusion	Major infusions of stored blood; incompatibility
Deep-venous thrombosis	Pyogenic or hydrostatic; contraceptive pill
Post-myocardial-infarction syndrome	Resorption of ischaemic muscle-breakdown products
Neoplastic disease	Disseminated neoplastic disease; leukaemia; invasive sarcoma
Recurrent pulmonary infarction	Usually associated with intracardiac thrombus
Crohn's disease	During an active phase
Sarcoidosis	During an active phase
Simulated pyrexia (thermometer cooking)	Attention-seeking or evasive
Trauma	Any major trauma with large areas of bruising, muscle-damage, haematoma, etc., even in out-patients; hyperpyrexia due to head-injury involving thermo-regulatory centres
Aspergillosis or other mycotic invasion	Highly specialized subject, needing hospital care
Histoplasmosis	A review article of endemic histoplasmosis in the USA, and of sporadic epidemic outbreaks of histoplasma pneumonia, gives valuable insight. The role of bird-droppings as a vehicle for the fungus and the importance of serological confirmation of the diagnosis are emphasized. Amphotericin B is said to be specific in treatment. Side-effects are of importance and the IV regime needs specialist supervision. Cutaneous histoplasmosis in Africa is caused by *H. capsulatum var. duboisii* – a very different fungus disease presenting as a subacute granuloma of skin or bone. Treatment as above

Addendum to p. 132a

Congo/Crimean haemorrhagic fever in Dubai (Suleiman and others, *Lancet* ii 939 (1980)) – a useful summary of a brief but alarming epidemic in a hospital with 3 fatalities out of 6 cases: tick transmitted; secondary spread prevented by barrier nursing; prophylactic antibiotic for secondary infection; cortico-steroids; fresh blood, and platelet concentrates for haemorrhagic phase; convalescent serum if available; no specific treatment

(*Cont. from p. 131*) which overwhelms people who think they have been exposed to infection is fully justified by the appalling nature of the 'furious' form of the disease. The French 'la rage' is a telling name.

It is important that every patient who presents with a possibility of a bite of a rabid animal should be treated seriously, and proper measures for protection and control taken. Foxes and vampire bats are the chief wild vectors.

The earliest symptom is paraesthesia at the site of the bite. Once hydrophobia and muscle-spasm have set in the prognosis is therapeutically hopeless but cases have been known to recover in intensive care as the virus ultimately loses its virulence.

The clinical steps are:

1   Careful and detailed history of biter and bit
2   Early and thorough irrigation, sterilization with tincture of iodine, and surgical excision of the wound
3   Infiltration of tissues proximal to the wound with hyper-immune human anti-rabies serum (HHARS) and IM injection of 10 ml
4   Admission for observation and continuing treatment with human diploid cell substrate vaccine (HDCSV)
5   Information of the relevant community health authority so that steps can be taken to capture and keep the biter for observation of its clinical progress

HHARS and HDCSV are available in Great Britain at Regional Public Health Laboratory Service (PHLS) laboratories. Enter the appropriate source on the facing page for your own area. The Colindale Laboratory (phone (01) 205 7041) is the headquarters of the PHLS.

The general control of enzootic rabies is outside the scope of this work, but it should be pointed out that the active immunization of all dogs in the communities most at risk produces a dramatic fall in the incidence of clinical rabies.

RADIOGRAPHERS

All radiographers are superb and thoroughly co-operative. If they are given requests which are specific and warranted by the clinical condition they are magnificent. Often there is a radiographer on

Rabies vaccine is available in the UK from Public Health Laboratory Services in Belfast, Birmingham, Cardiff, Colindale, Exeter, Leeds, Liverpool, and Newcastle, and from Hospital Pharmacies in Aberdeen, Royal Infirmary; Dundee, King's Cross; Edinburgh, Western General; Glasgow, Ruchill; and Inverness, Raigmore.

call after 5p.m. and on holidays, who very reasonably objects to being unnecessarily sent for: e.g. if you reduce a Colles's fracture it is fair to wait till the following morning for a post-reduction radiograph; and if a minor fracture needs X-raying on a holiday it is fair to wait till there is a small collection before summoning radiographic aid. An image intensifier usually fills the gaps, and ensures that you don't miss a significant fracture by misjudgement: this only applies to those accident units which have one and to those doctors who know how to use it.

RENAL DIALYSIS

Blocked arterio-venous shunts should be dealt with as follows:
1 Identify arterial and venous components
2 Clamp both with Blalock clips
3 Undo teflon union
4 Release arterial clip to verify patency, and replace
5 Suck out venous component after removal of clip
6 If no venous flow is established by 5, pass appropriate Portex nylon IV cannula (starting with 0.75 mm — French gauge 3) as far up the vein as possible and irrigate with normal saline containing 1000 units of Heparin to the litre; apply intermittent positive and negative pressure to the cannula by means of a 2 ml syringe and continue until exhausted
7 If no flow is established by these means return the patient to the hospital of origin or to the renal unit at the nearest relevant hospital

RENAL EMERGENCIES

See Uro-genital emergencies.

RESPIRATORY EMERGENCIES

Bronchial asthma   See Asthma.

Spontaneous pneumothorax   See Pneumothorax.

Mediastinal emphysema   See Chest injuries.

It is a professional obligation when ordering an X-radiograph to give the reporting radiologist specific clinical and anatomical details.

There is an obligation on Casualty Officers to open the door for kidney-donations in cases of fatal head-injury who are kept oxygenated on a ventilator. A final decision is seldom theirs, but they can be a useful starter.

Pneumonia or **bronchitis** in patients with pulmonary insufficiency (e.g. obstructive airways disease, pulmonary fibrosis, emphysema): urgent admission and 100% oxygen; by endotracheal tube if necessary; brief periods only.

Bronchiolitis in infants: oxygen tent and steam (see **Collapse and coma**).

Major haemoptysis   Suction, sedation and oxygen. Management often needs admission. (Generally neoplastic in developed countries, but more often tuberculous among poor and undernourished populations.)

Impacted foreign bodies   See **Collapse and coma**.

RESUSCITATION, EMERGENCY

1 Airway   If satisfactory, proceed to 2; if not, Guedel airway or endotracheal tube, and positive pressure ventilation.

2 Heart-beat   If satisfactory, proceed to 3; if not, electrical cardio-version, IV lignocaine, and medical transfer.

3 Blood-pressure   If satisfactory, proceed to 4; if not, restoration of circulating blood-volume, and transfer to specialist unit.

4 Diagnosis   If satisfactory, proceed to 5; if not, refer for specialist opinion or admission.

5 Treatment   If satisfactory, proceed to 6; if not, refer for specialist opinion or admission.

6 Disposal

Cf. **Cardiac arrest**, **Collapse and coma**, **Chest injuries**, **Shock**.

List of drugs to be kept available at all times on the resuscitation trolley

Hypoglycaemia
    Dextrose inj. 50% 20 ml
Cardiac and respiratory emergencies
    Sodium chloride inj. 0.9% 10 ml
    Water for inj. 10 ml
    Aminophylline 250 mg 5 ml
    Sodium bicarbonate 8.4% w/v 200 ml

Potassium chloride inj. 1.5 g in 10 ml
Adrenaline 1/10 000 10 ml
Adrenaline 1/1000 1 ml
Atropine sulphate 60 μg in 1 ml
Lanoxin 0.5 mg in 2 ml
Lasix 20 mg in 2 ml
Isoprenaline 2 mg in 2 ml
Calcium chloride inj. 13.4% 10 ml
Calcium gluconate inj. 10% 10 ml
'Xylocard' is a 20 mg/ml solution − dilute 1 ml in 100 ml of
   IV fluid
Lignocaine hydrochloride 5% 10 ml (500 mg in 1 litre of IV
   fluid)
Disopyramide 10 mg/ml
Mephentermine 15 mg in 1 ml

Epilepsy and head injury
Diazepam inj. 10 mg in 2 ml
Diazepam inj. 20 mg in 5 ml
Sulphamezathine 0.5 g in 3 ml
Dexamethasone 4 mg in 2 ml
Methylprednisolone 40 mg in 1 ml
Methylprednisolone 500 mg in 7.5 ml

Poisoning
Naloxone 0.4 mg in 1 ml
Doxapram 100 mg in 5 ml
Paraldehyde 5 ml
Amylnitrite (vitrellae)

Pain
Fortral inj. 60 mg in 1 ml
Fortral inj. 30 mg in 1 ml

Anaphylaxis
Hydrocortisone sodium succinate 100 mg
Depo-Medrone 4 mg in 2 ml
Adrenaline as above

RETENTION, ACUTE/CHRONIC, OF URINE

Acute retention is agonizing and needs early relief by catheterization. Always use a balloon-type of self-retaining catheter, leave it
in place attached to a urine bag or under water seal, and admit

Local variations and changes in fashion should be recorded here.

All α- and β-blockers and -unblockers best kept locked up in poisons cupboard.

Dopamine is given by IV infusion 800 mg in 500 ml of IV saline (*not* in bicarbonate). It is α-adrenergic and inotropic and comes in 5 ml ampoules containing 40 mg/ml.

Opiates locked in poisons cupboard by law in UK.

Major, multiple and chest injuries (q.v.)
   Methylprednisolone 2 g in powder form with solvent (Solu-Medrone) to be given IV fairly rapidly, via drip.

for observation. Chronic retention with overflow is best decompressed by stages so as to avoid reactionary cystic bleeding. Some urologists like to do this themselves after admission, but if pain is severe it is not kind to delay.

Preparation with IM or IV pethidine and local urethral analgesia with lignocaine-gel 2% is a source of comfort to the patient and makes catheterization more uniformly successful.

Regular catheterization of paraplegics sometimes falls to the lot of A & E staff. 3—4 weekly is optimum and transparent plastic better than rubber; silicone-coated best of all.

RINGS, FINGER

Casualty staff are experts with soap and persuasion. The ring cutter is easy and effective. Rings should *always* be removed from arm and hand fractures and infections of the same side.

SHEEP-TICK

Holiday-makers suffer from this infestation more · often than shepherds. It usually presents as a painless or slightly irritant pink purse, 0.5—1.0 cm in length, attached to the skin of the abdomen. The routine treatment is to press the blades of a fine curved artery-forceps on the skin either side of the thorax of the tick at the site of entry, spray the abdomen of the insect with ethyl-chloride or other cold spray to induce it to let go, close the forceps, and avulse sharply. The removal is generally complete and the death and sterilization of any residue is ensured by painting twice daily with tincture of iodine for 3 days. Tetanus toxoid antigen as required.

SHOCK, SURGICAL (reduced circulating blood-volume)

IV treatment in the accident department

It is important when making your decision as to what to infuse, and how much, to consider the following basic questions:

> **What is the aim of transfusion or infusion?**
> Is it to save life now?
> Is it to render the patient fit for surgery after transfer to another hospital or to a specialist unit?

Is it to gratify an irrational IV urge?

How much has the patient lost?
 Has he lost blood?
 Has he lost plasma?
 Is he still losing?

What does he need?
 Does he need simple fluid?
 Does he need plasma?
 Does he need blood-volume?
 Does he need blood?

It is useless to pour anything into a bleeding patient when blood-loss is continuing untreated. If blood-loss cannot be staunched in the accident department IV infusion may keep him alive until he can reach the theatre for this to be done under general anaesthetic. But, if bleeding is severe, it may need to be done with O rhesus-negative blood or plasma expanders (q.v.) before properly cross-matched blood is ready. Most A & E units carry a small stock of O rhesus-negative blood which can be used in extreme emergency while urgent cross-matching is awaited. If you work in a big unit which has its own theatres and a traumatic surgeon, your patient has the best chance of recovering. In smaller units the accident-department-to-surgeon shift often presents difficulties and delays, but it is of paramount importance to make it swiftly and effectively when the occasion demands. Try to use a micropore filter if you think that infusion of more than 2 units of blood is likely to be needed; the need for rapid infusion may prevent its use.

 It is entirely useless, not to say positively harmful, to continue pouring non-haemoglobin-containing fluids into an exsanguinating patient, as the result is only to produce dilution of the blood that remains. A drop of 70 per cent of circulating haemoglobin will be fatal in an acutely injured and bleeding patient who is otherwise healthy. Therefore give Haemaccel for urgent blood-volume replacement unless the loss is of plasma (e.g. in extensive burning) and get blood *urgently* to continue the infusion until the patient can be got to theatre. Remember that infusion of pooled plasma carries a 10 per cent risk of serum hepatitis. Human Plasma Protein Fraction BP is safer in this respect. If you use Dextran, take the blood sample for cross-matching before starting infusion.

If you use Haemaccel no such precaution is required. Cf. 'Plasma expanders'. Give 2 units of blood to 1 unit of Haemaccel.

Example

A patient with multiple injuries arrives in the accident department, having already lost 3 litres of blood from his circulating volume. Taking 6 litres as his estimated normal circulating volume, he has lost 50 per cent of his haemoglobin (i.e. oxygen-carrying) complement. If you then infuse 3 litres of non-blood of any kind you will have effectively reduced his haemoglobin by a further 50 per cent, i.e. to 25 per cent. If his bleeding continues and infusion of non-blood is maintained the haemoglobin will rapidly fall to levels incompatible with cerebral survival.

Conclusion

Infusion of up to 2 litres of Haemaccel in the accident department is a first-aid measure in the treatment of shock. If more than this is required it is essential to give blood. If there is continuing blood-loss which cannot be controlled, *instant transfer to the theatre is essential*.

SKIN EMERGENCIES

are few but skin conditions are commonly presented to A & E units. They are best recognized from experience in a dermatological clinic or GP's surgery. Illustrated textbooks are helpful but verbal descriptions not. Once the condition is recognized the patient should be passed on to the appropriate persons — dermatologist or general practitioner.

True emergencies

such as acute exfoliative dermatitis, pemphigus, and toxic epidermolysis seldom reach the A & E unit, but, if they do, need admission. They respond to parenteral steroids, but nearly always need nursing care and continuing supervision.

Acute exanthems (confluent chicken pox, haemorrhagic measles) need admission to isolation and are extremely toxic. They do not

Military antishock suit/trousers (MAST) An original pneumatic tamponade device for treatment of shock; not yet fully explored or validated, but popular in US as first aid. It consists of 3 segments, 1 for each leg and 1 for pelvis and abdomen. A development of the Newcastle G-suit. It offers help in restoring central blood-volume and controlling major pelvic bleeding in young victims of violent trauma, who often die of this. It does not replace hetero-transfusion and needs to be used according to the maker's instructions.

benefit from the primary treatment an A & E unit can provide, even when complicated by secondary infections such as pneumonia.

Minor emergencies

are listed below and mostly dealt with elsewhere in the text.

Cold injury (q.v.)

Dermatitis gangrenosa, *Erysipelas*, *Erysipeloid*: see **Infections, miscellaneous**.

Impetigo (q.v.)

Herpes simplex, *Herpes zoster*, *Paravaccinia*, *Orf*: see **Viruses**.

Sensitivity reactions (see Anaphylaxis, acute)

Contact dermatitis (e.g. pharmaco-dermatitis): metals, cement, chemical solvents. Cf. **Burns**.

Basic rules of primary care:

1 No water
2 Liquid paraffin for cleaning
3 Topical steroids
4 Refer to GP or dermatologist

Insect sensitivity (especially wasp — see **Stings, wasp and bee**)   IV steroids if severe.

Acute generalized urticaria can be very distressing and combined with abdominal pain in children. Antihistamines are palliative.

'Angio-neurotic' oedema — stupid name, frightening complaint. Mild sedation, IV steroids, 3—5 day course of anti-histamines, for both. Refer to GP. May involve the glottis. Cf. **Laryngotomy**.

Surgical

See **Hand, Grafts, Soft-tissue injuries**.

Replacement of large ablations of *senile or steroid skins* by traffic, domestic, and horticultural calamities often gives unexpectedly successful results and saves much distress. Always worth trying: treat as Thiersch grafts.

Tropical skin disease is a large and complicated subject in which local knowledge is of paramount importance. It is however worth mentioning cutaneous *Leishmaniasis*, which responds to a combination of general topical remedies together with systemic sodium stibo-gluconate, pyrimethamine, and sulphonamides. Cf. **Infestation, Yaws.** Antibiotics may be needed for secondary bacterial infection. Oriental sore − *L. tropica*; Espundia − *L. brasiliensis.*

Careful defatting of skin and careful debridement of recipient area essential. Admission for elevation needed.

Lumps and bumps

Most accident departments run a lumps and bumps service, as they are the best equipped to deal with them. They need out-patient treatment, patience, care, and follow-up. Any malignant or unusual skin-conditions should be transferred for specialist follow-up.

Acanthomata, Bowen's disease, epitheliomata, basal-cell carcinomata, naevi, various cysts, horns, fibromata, warts, and, alas, malignant melanomata all turn up from time to time. It is said that malignant melanoma has characteristic appearances but if so it is probably too late to do any good. Obviously it is best dealt with by experts, but if I get one I want it widely excised on the same day by anyone who is willing to do it, and not to wait till the week after next. Anyhow it may all too probably be identified only in the pathologist's cell. Cf. **Nails, digital**.

SOFT TISSUE INJURIES

Skin (see Wounds)

Good skin-repair using elementary plastic-surgical techniques up to specialist standard should be the aim in A & E units. Remember that infantile skin has an almost infinite capacity for regeneration and repair; the tougher the skin the tougher the road to recovery. Heavy manual workers' skins can be slow and difficult to heal, needing prolonged and painstaking after-care.

Muscle

Careful debridement — make sure that all dead tissue is excised and all external contaminants liberally slooshed out; then careful suture according to surgical principles and adequate support or splintage in relaxed position with early movement as the situation indicates. If serious injuries demand splintage for the joints above and below the repair, energetic movement of more remote joints is to be insisted on so as to maintain circulation and venous return.

Tendons

See **Tendon injuries**.

Lipomata are common and larger ones easily identified. Excision biopsy helps to separate Durcum's adiposis dolorosa from neuro-fibromatosis in its early stages.

Pretibial lacerations

Avoid stitching and use steristrips. Do not tie off minor bleeding vessels. Avoid haematoma by suitable elevation and foam-pad pressure dressings. All large flap-repairs need admission.

Distally based flaps can be carefully defatted and scraped so as to form thick Thiersch grafts or thin Wolfe grafts. Same post-operative care as above. They do well, even in senile and steroid skins.

Nerves

For repair of digital nerves see **Hand injuries**.

Repair of major nerves, e.g. median or ulnar at the wrist or radial at the elbow or mid-arm, is the job for the highly specialized traumatic surgeon who is prepared to use microsurgical techniques. If no such is available any experienced surgeon prepared to take pains must be found. This type of repair is in a 'growing region' of surgery and one must look forward to new advances all the time. Cf. also **Chest injuries, Abdominal injuries**.

STAB WOUNDS

These always need to be regarded with great suspicion. Three instances are given here of their latent dangers, and illustrate this point.

Stabs in the back

There may be an insignificant skin wound with no bleeding and the patient may complain of very little, but there may be a slowly accumulating haemothorax or haemopneumothorax which has no definite signs to start with. All need chest X-ray and observation, and a few need exploration. Cf. **Chest injuries**.

Stabs in the belly

The same caution applies. A negligible skin wound may be given by a small sharp instrument, and a probe may find no way through the sliding planes of the abdominal parietes, and yet there may be a small bowel perforation which could be fatal if it is unrecognized. An erect abdominal film may show air under the diaphragm, but even if it does not, admission for observation is essential, and exploration if there is any doubt about the extent of the injury.

Stabs in the groin

such as butchers often suffer when boning beef can produce lacerations of the femoral vein or its saphenous tributaries which become occluded by interstitial pressure for a few hours, only to break out again through normal walking movements disturbing the clot and altering local tissue relationships. Bleeding from this

Digital gangrene, blanket gangrene, **trichanchonē** (due to strangling of a digit by a hair or thread): see **Injuries, minor**.

Cf. **Abdominal injuries**.

area can be rapid and severe. Careful exploration of the whole wound is obligatory.

There are many variations on this theme (such as flying spicules of glass or metal) and they must always be borne in mind. A dramatic example of an unexpected outcome is afforded by the history of the little boy brought in with acute distress and cerebral disturbance, but with no external stigma apart from a tiny wound in his right upper eyelid. In the event it became clear that he had been playing in the garden with the broken blade of a miniature hacksaw, had fallen forwards on it in such a way that it passed through the orbit (missing the eye), through the superior orbital fissure and into the brain, penetrating as far back as the thalamus on the same side. He died about 10 hours later after coma, fits, and hyperthermia. The blade was found later in the garden with his blood on it. He must have pulled it out himself an instant after wounding.

STINGS ETC. BY MARINE CREATURES

Conan Doyle gave the classic description of fatal injury by the polyp *Cyanea capillata* in 'The Lion's Mane' (*The Case-Book of Sherlock Holmes*), and deaths due to stinging creatures in the sea (especially among children in tropical waters) still occur. If early treatment can be given with IV anti-histamines and topical and systemic steroids, in anaphylactoid types of reaction with IV adrenaline, lives can be saved. General supportive treatment is needed as well. Cf. **Stings, wasp and bee, Anaphylaxis, acute,** and **Collapse and coma.**

Other examples of stinging creatures are jelly-fish (e.g. *Aurelia aurita*) and the dangerous Mediterranean *Physalia* (Portuguese man-of-war), weaver fish, sting-ray, and other stingers in tropical waters. The unfriendly habit of mediterranean sea-urchins of leaving their spines in bathers' soles is well known. These produce an intensely painful foreign-body reaction and, even after careful removal, a persistent area of sensitivity and inflammation which may need topical steroid treatment, with occlusion, to obtain relief. The reaction can last for some months.

Glass wounds should always be regarded as deeply penetrating until it is proved otherwise.

A recent article on serious hand injury from stinging by the 'sea-wasp' (*Chironex*) jelly-fish is instructive. Early treatment with anti-histamines, steroids, and, in cases with extreme swelling, surgical decompression may be crucial (Drury and others, *Injury* **12** 66–8 (1980)).

Immediate soaking in acetone is said to dissolve the spines. Jamaican fishermen recommend rum.

STINGS, WASP AND BEE

A vast number of people attends A & E departments annually for this tiresome affliction without any necessity. It is simplest to issue a 3-day course of chlorpheniramine (Piriton) tablets as a routine. *But* a few come who suffer from severe local or general reactions. These need IM or IV Piriton 10 mg at once, and a 3-day course of tablets. Patients who show anaphylactic reactions with tachycardia and collapse need IV adrenaline 0.1% 0.1 ml each minute (or 0.01% 1 ml each minute), and occasionally IV fluids and admission. They should *always* receive a course of desensitizing injections against wasp/bee venom. IV hydrocortisone 100 mg or dexamethasone 5 mg is worthwhile in severely collapsed cases. Cf. **Anaphylaxis, acute.**

SUICIDE, ATTEMPTED

No longer a crime. Usually a demonstration of withdrawal from an emotionally intolerable situation or a demand for attention rather than a determined attempt to 'end it all'. Every case should be admitted to a medical ward if treatment is needed for intoxication, to a psychiatric ward if not. A three-day order may be necessary. This can be procured from the police or from the relevant social worker (formerly the Mental Welfare Officer). See **Insane, certification of the.**

In my experience the suicides among children under 14 (especially boys) and adults over 60 (especially men) are usually serious attempts. It is fashionable to attribute these to depression (in the technical sense), but often, in fact, the children, from deep affection, feel that life after parents are divorced or separated is intolerable, and the elderly, from loneliness, despair, or disease, make the determined effort to be rid of what they cannot bear.

SUTURE MATERIALS

The choice of these presents a difficult problem with as many answers as practitioners. In Hereford we used a well-known synthetic polyglycolic absorbable suture, Dexon, exclusively for

Bee stings

Anaphylactic reactions may put the victim's life in jeopardy. A review article (Frankland and Lessof, *J. R. Soc. Med.* **73** 807 (1980)) puts the matter into admirable perspective. Venom immunotherapy (specific hyposensitizing material currently available from both Pharmalgia and Dome) is there and has been shown to be effective. It needs to be given in a hospital environment, at least to start with, under observation by an experienced person able to deal with the earliest signs of anaphylaxis. Not to give it when a patient has experienced an anaphylactic reaction is hardly justifiable (cf. *Lancet* **ii** 956 (1980)). A six-hour hyposensitization course is described by van der Zwan and others, *BMJ* **287** 1329 (1983).

2 years (1972–3) in the hope that it would prove time-saving, effective, and simple, but our hopes have not been adequately fulfilled, because of the irregular way it is resorbed by the patient's tissues. It may linger on for 4–5 weeks and require as much supervision an an unabsorbable material. It is not possible to say 'You have been repaired with Dexon: your stitches will disappear on their own with no further attention.'

· The crux of using Dexon comes in facial repairs, where the problem is as follows: multiple facial lacerations, so common in RTAs, can be sutured satisfactorily with fine non-absorbable sutures provided they are removed about the fourth or fifth day, or with 5/0 Dexon provided it is carefully supervised. In the first case it may be impossible or very painful to remove dozens of fine sutures at the fifth day, because of the universal scabbing associated with this type of injury. If 5/0 Dexon is used and lingers too long it may require repeated follow-up visits to ensure that all is well, and if a patient is transferred to another hospital this need may not be fully appreciated. Scabs usually separate from the second to third week and tidying up can be done then. Dexon is unpopular with plastic surgeons because of its tendency to produce thickened scars. Otherwise it can well be used for hands, scalps, children whenever possible, and some extensor tendon repairs. Its advantages in these fields outweigh its disadvantages.

For facial wounds it is wisest to use a fine non-absorbable monofilament material and remove it early if possible. The fine monofilament synthetic sutures can be left for surprisingly long periods without producing any reaction. Braided silk has no place in any form of repair in A & E work. It can produce a foreign-protein reaction and is a potent source of secondary infection because of its woven capillary structure.

TENDON INJURIES, MISCELLANEOUS

Biceps brachii rupture

Mostly in elderly men and spontaneous, occurring in the proximal tendon of the long head of biceps over the head of the humerus. They are not repaired and do well. Traumatic division at the

common tendon of insertion is repaired on general surgical principles and does well.

Abductor + extensor pollicis longus

Whether divided singly or together, these need stainless steel wire and abduction splintage in working men. In elderly ladies after Colles's fractures, rupture of extensor pollicis longus is best left alone. Some repair it. See **Fractures**, p. 65.

Quadriceps femoris tendon (supra- and infra-patellar)

Often associated with injury to the joint capsule. Very much an orthopaedic ploy.

Achilles tendon

Historically characterized by the feeling as of a blow on the calf or below it, while running or jumping, and usually accompanied by a fall. Most orthopaedic surgeons prefer primary repair followed by an equinus plaster. Admit. Cf. plantaris syndrome, p. 52.

TENDON REPAIRS, HAND

Extensor tendons

There is little disagreement about the primary repair of *extensor tendons* in the A & E department. It should be a routine procedure in which all senior and some junior staff are proficient. The conditions provided should be of operating-theatre standard, as far as it is possible to achieve this in A & E circumstances, and the steps approximately as follows:

 1 Extensive skin preparation, including trimming and scrubbing of the patient's finger-nails.

 2 Scrupulous cleaning and debridement of skin wounds with careful assessment of available skin-flaps. Mobilization of existing flaps gives much more effective skin cover and should be done whenever feasible — however awkward. Grafting is a last resort, but better than no skin.

 3 Dissection of two ends of a completely divided tendon for at least 1.0 cm above and below. Where the tendon is expanded it should be united by Dexon 3/0 or Ethibond stitches — interrupted

4/0 gauge on a round-bodied needle. Where the tendon is cylindrical an initial stitch (Kleinert's method preferred: see Fig. 22) using a suture with a needle on both ends is required, and interrupted stitches afterwards. This gives a secure hold without interfering with the blood-supply. Dexon (see **Suture materials**) is said to be inert in the tissues and absorbed in about 3–4 weeks, thus fulfilling all the requirements for extensor tendon repair. Silk is an animal protein producing foreign protein reaction; non-absorbable synthetics act as permanent foreign bodies. If a repaired tendon is under unacceptable tension a single stainless-steel wire passed through the tendon, well proximal to the suture line, and brought out through the skin distal to it, where it is joined over the appropriate button, gives security. It produces no reaction and can be painlessly removed at the tenth day.

4 Skin closure with 4/0 Prolene, carried out with the greatest attention to detail in obtaining good apposition while preserving blood-supply.

5 Minimum dressing applied with synthetic occlusive glue (Nobecutane, Opsite, etc.). No dressing but glue after the third or fourth day.

6 Minimal but logical extension-splintage. Thus if the injury is (a) on the dorsum of the hand or at the metacarpal joint, the wrist should be fixed in extension with POP from lower third of forearm to distal phalanx; (b) distal to the metacarpal joint and proximal to the distal interphalangeal joint, Zimmer-splintage

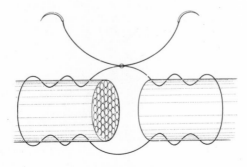

Fig. 22.

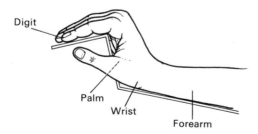

Fig. 22a. Volar slab POP for extensor tendon injuries and hand fractures (especially metacarpal and proximal phalanx). Metacarpophalangeal joint at 90°; wrist joint at 60°. Bandage to metacarpophalangeal joint only.

(padded, malleable aluminium strip) from proximal to the meta-carpal joint to distal to the proximal interphalangeal joint; (c) distal to (b), splint from finger-tip to middle of proximal phalanx. Splints should be as small as is possible to do the job, so as to allow as much movement as possible without putting stress on the suture line. They are only required for 10–14 days for uncompli-cated extensor-tendon repairs. The results are good – often better than you would think possible.

In complicated, multiple, or comminuted injuries of the hand and fingers, when all possible repairing has been done, a volar slab gives the best support. It should extend from elbow to finger tips, ideally with the metacarpophalangeal joints at 90° and the interphalangeal joints straight, but some flexion combined with movement from the earliest possible moment suffices.

Cf. **Hand injuries, Fractures,** Hand.

Flexor tendons

Flexor tendons in the hand are often cut and provide in their repair one of the most difficult and contentious problems in surgery. If the tendon is divided in the danger area (from the carpal tunnel to the distal interphalangeal joint) repair may be comparatively simple, but restoration of function is bedevilled by irresistible adhesions between the repaired tendon and the adjacent tissues, making the whole undertaking useless. This applies especially in the fingers.*

This is not the place nor this the writer to describe methods by which this disastrous state of affairs can be overcome, but it is my firm conviction and experience that in a majority of cases primary repair of flexor tendons in this region is not only possible but obligatory, and the results vastly better than secondary repair and tendon grafting in the hands of any but a very few outstanding experts. A successful primary repair should produce a normally functioning finger in 6–8 weeks. Secondary repair and grafting can dawdle on for months or years with results often short of satisfactory. This alone makes the perfection of techniques of primary repair of unequalled urgency and value to the injured. The key components of success in my experience are:

Early operation – within 4 hours of injury

Scrupulous techniques of tendon suture

*In the region overlying the proximal phalanx.

Resection of the fibrous flexor sheath over the area of tendon repair

Early mobilization in carefully designed splints, which allow active extension in combination with elastically controlled return (see Fig. 23)

Close and frequent follow-up, with relevant adjustment being made as often as the situation calls for it.

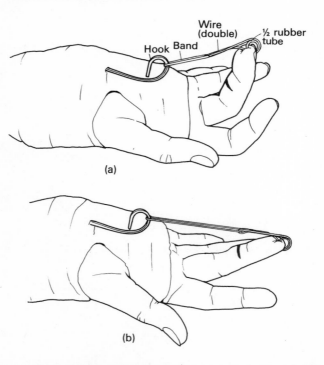

Fig. 23. Elastic splintage for repaired flexor tendons — the black line suggests the site of the injury. (a) Resting position maintained by elastic recoil. (b) Extended position. The hook is thick wire; the band is square rubber the fine wire is the finest stainless steel — a double loop through nail and skin; the horseshoe is a $\frac{1}{4}$ in segment of a $\frac{1}{8}$ in diameter rubber tube. The POP is applied to the flexed wrist (about 30° of flexion).

Recent work suggests that the sheath should be repaired carefully and in detail and suggests other modifications in the programme outlined here. It is too early to recast the whole approach unless you are dealing with very large numbers of tendon repairs (Richards, *Injury* **12** 1–12 (1980)).

As the performance of this type of repair is so fraught with pitfalls it should be done by the most skilled and experienced only, but if such a person is not available it is best undertaken by someone who is available and who will make up for lack of skill by infinite patience and pains. A stainless-steel suture is good; braided polyester (Ethibond) may be better. Experience develops practice.

This situation illustrates clearly one of the few major problems of providing an adequate hospital service for the injured. Acute trauma often calls for the most expert and specialized care, which is the very care that is not available except in selected centres of excellence in large cities. The only approach to meeting this need in places far away from such centres (i.e., in a majority of accident units) is by intelligent and open-minded collaboration between all the available people on the spot.

Some difficult problems — severe head-injuries, chest injuries, lacerations of the heart and large vessels — are comparable: they are of vital importance and it is difficult to provide the best treatment in every case. But the effort has to be made and by rational teamwork a lot can be done to alleviate injuries which otherwise would be denied the care they need.

TENOSYNOVITIS

Common cause of pain in forearm, often with swelling over affected tendons. Can occur in any long tendons with tendon sheaths. Crepitus not constant. Treatment by POP, 10 days to 3 weeks according to the severity; concurrent anti-inflammatory drugs should be used; then injection of lignocaine and Depo-Medrone 2 ml if POP is not successful. This is not the ideal primary treatment as it has its own perils and may lead to too early reuse of the affected limb, with obstinate recurrence of the disorder. I prefer to use it as a secondary remedy. Remember to apply the plaster with due regard to effective immobilization of the affected tendon. A Colles's type of plaster (e.g.) is of no use to immobilize an inflamed extensor pollicis longus tendon.

Two other related complaints commonly presenting in A & E units are:

de Quervain's tenovaginitis stenosans

of the tendons of extensor pollicis longus and/or abductor pollicis, in the fibro-osseous canal at the radial styloid. No palliative measure seems any use and the canal has to be surgically explored and laid open.

Tenovaginitis stenosans

of the flexor pollicis longus tendon at the metacarpophalangeal joint presents in little boys (usually under 4 years) as a fixed flexion deformity at the interphalangeal joint. It is often bilateral and the tender nodule responsible for it is palpable at the metacarpophalangeal joint. This condition is responsive to surgery.

In middle-aged women (usually over 50 years) it generally presents as a painful, clicking, sometimes briefly locking, abnormality of thumb flexion. In other respects it appears identical to the childish version and in need of similar surgery.

TETANUS PROPHYLAXIS

Anti-tetanic serum

is prepared from horse-serum. It is dangerous and never to be used when human antibody is available.

Tetanus toxoid antigen

is safe and gives effective protection if administered within 24 hours of injury in previously immunized patients. If injection is delayed beyond 24 hours it is still worth starting a course of three injections or giving a booster. Tetanus toxoid antigen in simple solution seems to give less unpleasant local reactions than the adsorbed kind. If someone has had a full course of tetanus inoculation at *any* time in the past, a single booster only is required after injury. Because of higher antibody response primary inoculation is always with *adsorbed tetanus toxoid antigen* 0.5 ml. Secondary inoculation is adequate with *simple solution of tetanus toxoid antigen* 0.5 ml (if available).

Tetanus immune-globulin

Wellcome (1973) procured an anti-toxin prepared from pooled

Suppurative tenosynovitis
See pp. 83—4.

Both kinds are best done under local anaesthesia if appropriate. This allows the active co-operation of the patient and assures adequate release of the constriction.

TETANUS, DIAGNOSIS OF

This can be difficult: uncomfortable stiffness of the neck and jaw is the usual presentation, with vague malaise. The causative wound may be trivial or even unnoticed. There is still no immuno-logical means of identification.

The problem of *reactions to tetanus toxoid antigen* is dealt with in an annotation by Grist, *BMJ* **284** 1453 (1982). He advises the use of 0.1 ml of simple antigen intradermally in cases of doubt; this will not only give an indication of sensitivity by producing a skin reaction, but also an adequate antigenic stimulus with negligible risk of provoking anaphylaxis.

human plasma (Humotet). It is screened for syphilis and Australia antigen. It is desirable to use this in cases of injury where there has been no previous active immunization, or where circumstances make it impossible to verify. Active immunization should be started concurrently, at a separate site of injection. Prophylactic dose: 1 ml in an adult, 0.5 ml in a child. Regular use during the years since its introduction has not disclosed any undesirable side-effects. Humotet is of course superior to equine anti-tetanic serum in the treatment of established tetanus. I think it would be difficult to convince a court of law that a patient dying of tetanus without benefit of human antibody had been adequately treated.

It is worth recording that in the author's experience of clinically established tetanus, all six cases occurred after minor injury — a pinched finger in a child, a firework burn on the chest of a teenage boy, a scratch-wound on the foot in a young man, etc. No injury is too trivial to require proper immunization. There is an increasing number of recorded cases of clinical tetanus following episodes other than wounds (e.g. oesophagoscopy, inhaled foreign body, foreign body in the eye, simple ligation of piles), and it is familiar that clinical tetanus in veterinary practice is transmitted by droplet infection. In parts of West Africa it is one of the commonest infectious diseases, especially in the new-born.

The only rational policy in an A & E unit is to aim at universal immunization. Every patient attending should be offered immunization, every injured patient should be recommended to accept it. Documentation should be universal and thorough. Re-inoculation is advised every 10 years, and after 5 years if there is a new injury, unless contra-indicated by sensitivity reaction.

THORACIC INTUBATION

In cases of chest injury in which you suspect tension pneumothorax or haemo(pneumo)thorax, intercostal intubation may be life-saving in the early stages of reception.

Identify tension pneumothorax by passing a wide-bore needle between the second and third ribs in the mid-clavicular line. If air audibly escapes intubate and connect the tube to an under-

Established tetanus is, of course, the classic disease for treatment in an ITU. Every suspected case is to be admitted at once.

A rational scheme, now generally accepted, is described by O'Brien (*A & E News* **25** 8 (1982)), with supporting immunological studies.

water seal or one-way valve system. If common sense, examination, and radiography suggest that there is an accumulating haemothorax, intubation will materially assist pulmonary expansion, and so cerebral oxygenation, in addition to giving a useful indication of intrathoracic blood-loss.

Intubation is best done with a trocar-mounted thoracic tube and always at the second intercostal space in the mid-clavicular line. The skin should be incised by means of scalpel cut 1—2 cm in length before inserting the trocar. The apparatus should be grasped in the gloved hand like a dagger held 5 cm from its tip. This will avoid too deep a penetration. Rapid attachment to a previously prepared underwater seal follows. The tube can then be gently passed into the chest for a convenient distance and anchored to the skin with a stitch. If you use an axillary or posterior approach you run the risk of perforating a liver, spleen, or bowel which has wandered into the chest through a tear in the diaphragm.

Cf. **Chest injuries.**

TORTICOLLIS, ACUTE

This is characterized by onset at night. The young patient wakes with a painful wry-neck. X-ray not needed. It can also occur during exertion and is common in the winter, less common in the summer. X-ray is advisable in patients over 25 years.

Treatment

Spray skin over painful area with *local* pain-relieving spray and apply axial traction for 2 minutes, then slow, forced rotation to the unaffected side. If relief is not complete apply soft collar. Cure within 48 hours is usual, within 24 hours likely, within one hour frequent. Analgesics if necessary. Very severe cases may need general anaesthesia or IV narcosis.

TRACHEOSTOMY

Tracheostomy should never be undertaken as an emergency procedure, for the following reasons:

Distinguish from cervical PID by absence of root-signs, and by age-group.

It is not easy to do, if it is to be done properly, as the trachea is a long way from the skin at the desirable level.

Properly placed tracheostomy involves division of the thyroid isthmus, and in an asphyxiated patient the inevitable bleeding is likely to lead to contamination of the airway at least, and blockage of the trachea at most.

Improperly placed tracheostomy is likely to need refashioning (with all the attendant complications) if required for a long time. If it is not wanted for more than a day or so the operation is unnecessary.

It takes too long to meet an emergency.

Therefore in cases of asphyxia which cannot be overcome by intubation, laryngotomy (q.v.) — also known as crico-thyroido-tomy and laryngostomy — is the emergency operation of choice.

ULCERS, GRAVITATIONAL*

The chronic ulcers which used to provide material for a weekly clinic in many old casualty departments were more remarkable for their foul smell than for the effectiveness of any remedy supplied. Mercifully they exist no more and should never return, because of an improved understanding of the role of compression treatment of gravitational ischaemia. In the acute phase they should be treated with sterile foam pads soaked in an antiseptic solution (e.g. povidone iodine 10% in water), kept in position by elastic bandages or Tubigrip and encouraged by regular walking exercise. These need changing every two or three days until infection is controlled and healing has begun. Then they should be replaced by weekly application of Ichthopaste bandages expertly applied until the ulcers are healed. Then it must be made clear that appropriate elastic support is essential for evermore unless surgery (e.g. varicose vein stripping) can remedy the underlying cause.

If you meet an ulcer which does not respond to this approach it may be that arterial ischaemia is responsible, or else that you ought to have done a Wassermann reaction. Tertiary syphilis can still occur.

There is a wide variety of local applications available for controlling infection and hastening epithelialization. Ensuring adequate venous return is the basic and permanent need.

TUBERCULOSIS

Three presentations can occur as emergencies:

Acute pulmonary (miliary tuberculosis)
Not a cachectic presentation but a subacute broncho-pneumonia with fever and characteristic radiological features.

Glandular
Not uncommon in cervical glands in children, presenting as single, or groups of, rubbery glands diagnosed by excision biopsies. In the elderly as chronic, caseating cervical abscess.

Bone and joint tuberculosis
Common in Afro-Asian immigrants to UK. Spinal cases present with back-ache, malaise and loss of weight. Other bone and joint manifestations. Early identification very important.

Also common in poor or malnourished communities world-wide.

*These are properly the province of dermatologists.

Various manufacturers supply graduated elastic supports with a wide range of sizes (e.g. Tubigrip SSB). These are better than plain tubular supports. Cf. pretibial lacerations, p. 143a.

UNCONSCIOUSNESS

Causes

Self poisoning: history and absence of physical signs (pp. 123 ff.)

Cerebro-vascular accident (p. 112)

Epilepsy: history and scars (pp. 38, 41)

Diabetes: instant diagnosis with BM Stix (p. 44)

Cardio-vascular collapse (pp. 29, 32)

Febrile convulsions: see **Convulsions, febrile**, p. 41

Trauma: usually self-evident, but often associated with an alcoholic component (p. 49)

Rare causes   Miscellaneous, such as:

Uraemia (pp. 136, 158)

Addisonian crisis (cf. p. 53)

Anaphylactic shock (N.B. especially wasp-sting: see **Stings, wasp and bee**, p. 146)

Waterhouse—Friederichsen syndrome and various others too rare to merit inclusion (p. 39)

Cf. **Collapse and coma.**

URETHRAL INJURY, SUSPECTED

It is very important to look for this injury, especially in young victims of pelvic fractures from RTAs or from being rolled on by their fallen horses. Genito-urinary surgeons vary in their approach and you have to know what your own specialist prefers. If catheterization is required it should be done gently with a soft rubber catheter of the smallest gauge and only on direct instruction by the surgeon consulted. You must avoid increasing bleeding or making false passages. Some surgeons prefer to have a 'hypaque' urethrogram done as the only primary procedure, and to take over the subsequent care themselves. Of course this may not be feasible in a major accident when the patient has multiple injuries. This situation emphasizes the need for teamwork in severe and multiple injuries.

Inability to pass urine, haematuria, and dysuria are the presenting symptoms.

Severe urethral injuries are often associated with perineal bruising.
   Subcutaneous extravasation of urine or intra-peritoneal bladder rupture (both rare) are potent causes of a toxic type of shock.

URO-GENITAL EMERGENCIES

Anuria

Primary anuria is a medical emergency unlikely to arise in the A & E department; secondary anuria can be due to hypovolaemic shock (see **Shock**) or the crush syndrome. This is a syndrome identified in the First World War and forgotten, re-identified in the Second World War, and forgotten again until the Moorgate tube-train disaster in 1977 when it made itself remembered again. It should never be forgotten; it is caused by the blockage of renal tubules by products of myoglobin breakdown, which follows the release of the circulatory return from areas of prolonged muscle ischaemia such as crushing entrapment of a limb. It is the *only* condition which requires a tourniquet as first-aid treatment and emergency amputation as the definitive treatment. It should not be confused with renal shut-down. Cf. **Urethral injury** below.

Renal colic

Familiar to every house-officer; may be accompanied by haematuria. A common entrée for drug addicts as it is easily simulated and difficult to disprove. Abdominal X-ray may show a stone.

Epididymo-orchitis

should never be diagnosed in young men except in the presence of active urinary or venereal infection; likewise orchitis should never be diagnosed in anyone at all except in the presence of mumps. In young men and boys every such case is a testicular torsion unless proved otherwise — that is by operation.

Gynaecological emergencies (q.v.)

Haematospermia

Not uncommon in young men and generally of no import. It occurs with seminal emission and is generally noticed in night-wear. It can accompany new growths such as seminoma testis, but is a late sign. Associated with schistosomiasis in endemic areas.

Haematuria

May be associated with great fear but seldom requires urgent treatment. The only massive haematuria occurring as an emergency in the author's experience followed retrograde pyelography and arose as a sensitivity reaction to the contrast medium. It can follow IV pyelography similarly and as a late complication of prostatic cancer. Investigation of haematuria sometimes requires admission; clot-retention must always be admitted.

Oliguria

as a complication of injury is a manifestation of hypovolaemic shock. Every victim of major or multiple injuries should be catheterized (indwelling balloon-catheter) and the output monitored as a routine.

Paraphimosis (q.v.)

Phimosis

seldom produces total retention. Formal circumcision as an emergency or elective procedure is better than stretching. In the aged a dorsal slit may be kinder and as effective (see **Paraphimosis**).

Retention of urine

Catheterize with an indwelling balloon-catheter. If this is impossible, dilatation of a stricture may be needed — use urethral bougies. This needs practice, but the art can be acquired readily. In remote areas every doctor dealing with emergencies should be able to cope. Receiving surgeons fall into 3 classes: those who say 'How dare you lay your filthy hands on my patients?' (reply: 'To relieve their pain'); those who say 'Why don't you put in a catheter instead of bothering me?' (reply: 'To avoid the first response'); and those who use both gambits alternatively. There is a small class who respond rationally; they should be cultivated as assiduously as a vegetable garden in a famine.

Bladders suffering from chronic retention with overflow need decompressing in stages to avoid reactive bleeding from the kidneys and bladder wall. Cf. **Obstetric emergencies**.

Stone in the bladder

Seldom a source of emergency, but it can present as severe intermittent pain related to posture and can interfere with micturition. Commoner in men than women, in hot dry climates than in temperate, in poor communities than rich.

Testicular torsion

Never diagnose orchitis in young men and boys except in the presence of mumps. A tender or painful testicle must always be regarded as due to torsion until it is proved otherwise. This is a surgical emergency which is too often missed. It may resolve spontaneously but in a high proportion of untreated cases the poor chap is left with one ineffective gonad. Every case should be referred to the surgical specialist. Cf. p. 158, Epididymo-orchitis.

VENEREAL DISEASE

Suggestions from a consultant venereologist:

Early cases are not to be treated but referred

Florid cases with urethral discharge:
  1   Take plain swab into Trichomonas transport medium
  2   Take charcoal swab into Stuart's transport medium
  3   Take blood for Wassermann reaction and gonococcal fixation test ('VD serology' is a safe omnibus request)
  4   Send all three urgently to the lab
  5   Give 1.2 mega units of IM penicillin. Alternatively give Augmentin by mouth, 6 g in 24 hours
  6   Refer to next VD clinic at the appropriate time and place

VIOLENCE

Unfortunately an increasing source of injuries and of management problems in accident units. General violence (varieties of mob-hysteria and riot) is common and produces a wide variety of injury including smothering and unspecified assaults; individual violence, often associated with drunkenness, is commoner still at dances and outside pubs at closing time. The currently accepted form (1980) is for drunken idiot A to knock down drunken idiot B who is incapable of rising to retaliate; then A kicks B repeatedly

Dry slides for gram-staining should accompany the swab specimens. A large variety of other infections are currently recognized as sexually transmitted, e.g. *chlamydia*, *trichomonas*, *candida albicans*, *gardnerella*, hepatitis A & B. Some types of vaginitis and metritis can still be legitimately described as 'non-specific'.

Acquired immune-deficiency syndrome is a new, frightening, and unnecessary disease with no treatment currently available. Usually fatal. (Review article: Waterson, *BMJ* **286** 743 (1983).) Mostly male homosexuals. Association with Kaposi's sarcoma suggests hope of treatment as well as oncological illumination.

Genital herpes Referral for identification and treatment in all cases.

in the face and trunk. Maxillo-facial injuries are common and often severe. Associated skin wounds are usually contused and difficult to repair.

A great and growing social problem in the West, reflected at all levels of society with varying degrees of sophistication; chiefly verbal still at the professional level.

Management of violent patients leans heavily for help on the police, who are very good at coping. This is another reason for maintaining good relationships with the force — within the bounds of confidentiality.

VIRUSES

Acute virus infections often arrive in the A & E department because symptoms may come suddenly and manifestations may be unfamiliar or confusing, and attributed to trauma. The briefest survey, with briefer first-time treatment, is all that is required.

Herpes

Simplex (*labial* (cold sore); *digital* (herpetic whitlow); *genital*) Effectively treated with idoxuridine 5% in dimethysulphoxide vehicle. This is applied as local compress firmly for 15 minutes or rubbed into the lesion with a soaked cotton-wool pledget mounted on a wooden applicator. If it is applied for longer periods it tends to macerate the surrounding skin.

Herpes simplex keratitis (dendritic ulcer) is treated with ophthalmic idoxuridine, but needs ophthalmological supervision in case of complications.

Ophthalmicus (Zoster virus) Early administration of adrenocorticotrophic hormone (Acthar 40 units or Synacthen 2 mg, twice at 48-hour intervals) will usually abort an attack. This treatment is disputed by some. Be guided by your own ophthalmologist. Seal with glue (Nobecutane). Ocular involvement is generally a later complication and needs expert ophthalmological supervision.

Zoster Local lesions respond to idoxuridine, as above, and an attack diagnosed early can often be aborted similarly. In the elderly, in whom post-herpetic neuralgia is such a dire consequence, the opportunity is never to be missed.

Bell's palsy

may be due to a herpetic virus. I am convinced by experience of the usefulness of the adreno-corticotrophic hormone treatment for this condition too. It is probably of varied origin, so a variable response is to be looked for.

Aphthous stomatitis

Idoxuridine as above, applied by local pressure of a well-soaked pledget to each patch for 4 minutes twice daily.

Paravaccinia (milker's nodule) (cf. Orf)

Generally a digital and uncomfortable lump like a typical primary vaccinia. Differentiate from the following by history of contact with the relevant host. Fairly rapid onset. Idoxuridine apparently specific.

Orf (q.v.)

Cat-scratch fever and tularaemia

Cat-scratch fever is clinically similar to tularaemia and differentiated fully only by antigenic studies. It is of viral origin (probably a chlamydia).

Tularaemia is a bacterial infection. The primary lesion is an indolent ulcer of the hand or finger as a rule, due to a bite from an animal carrying *Yersinia tularensis*, with axillary or cervical rubbery glands which may break down to a sluggish and barely painful abscess. Biopsy of a gland shows tuberculoid granulomata with giant-cell systems, but no acid-fast baccilli. In both cases treatment is expectant, with active attack on secondary infection.

WOUNDS (cf. Soft tissue injuries)

The general care of wounds is extremely important and not always made enough of. An A & E department with a bad record of sepsis is one in which there is not a high enough standard of wound-care. Antibiotics and antiseptics are not good substitutes.

Finger wounds should be cleaned to the wrist, hand wounds to the elbow, forearm wounds to the shoulder (above and below),

Or 1% Idoxuridine mouth-wash (contrast *Behçet's syndrome*, in which oral ulceration may precede uveitis and skin, joint, and CNS symptoms by months or years: treated with topical/systemic steroids).

It is important to recall the increase in the incidence (and understanding) of viral diseases in recent years. This is a field in which the world of A & E medicine is not often primarily involved. Further clinical information in rare instances is generally available from dermatologists and venereologists. Virological specialists are now widely available in pathological laboratories.

and so on. Skin edges should be trimmed to make them neat and straight; remove epidermal tags, any devitalized tissue and all visible foreign bodies; irrigate invisible foreign bodies with detergent solutions (e.g. cetrimide 1%), get rid of grease with detergent jelly (e.g. Swarfega), dry the wound carefully, and only then repair. You cannot take too great pains at this task.

Suture should be carried out carefully and thoroughly so as to secure perfect apposition of skin edges without slackness, irregularity, or tension. The perfect suture line obtained should be sealed with occlusive glue and given the minimum of dressing and the maximum of ventilation. It will then be painless, clean, and quick to heal in ninety-nine cases out of a hundred. If primary closure is not possible, admission for subsequent grafting should be considered. Every hand wound should have a sling for 48 hours at least. After-care of skin wounds *after* healing by application of surgical spirit after washing, or chlorhexidine cream to scabby wounds, reduces secondary infection to vanishing point.

Antibiotic sprays should never be used: they fill the air with allergogenic dust, and the department with multi-resistant bacteria. Dry povidone—iodine sprays are good for antiseptic application, and wet ditto equally, but not so convenient.

If you are making the wound yourself never forget that the skin you mean to cut must always be cleaned with a detergent first (soap and water or cetrimide 1%), dried, and then well painted with tincture of iodine (weak solution of iodine in spirit BP). Sensitivity to this preparation is rare, and reversible with steroid creams.

WRIST AND HAND AMPUTATIONS

and wounds involving the main nerves, tendons, and blood vessels are of increasing importance as specialist techniques for dealing with them improve. The use of microsurgical repair and its application to replantation operations offers new hope in these dreaded injuries.

If it is decided to refer a patient to a plastic surgery centre for this type of operation the detached finger or hand should be wrapped in sterile towelling and enclosed in a thermo-stable container holding ice, but not allowed contact with it. Only

Lacerations of senile and steroid skins: see p. 143a.

Powders are only active in solution − i.e. moist wounds only.

clean-cut wounds are likely to be amenable to this treatment (e.g. guillotine amputations). The longest possible warning should be given to the receiving unit as advance notice of an impending emergency operation which may take from 12 to 18 hours is obviously needed.

X-RAYS

are dangerous and expensive. Requests for examination and films should be limited and specific. Do not X-ray fractured ribs unless there is a positive indication, e.g. lung injury, nerve injury, or anticipated litigation. Never ask for an X-ray without detailed examination and *then* make a specific request. For example if you suspect a fractured scaphoid it is useless to ask for 'X-ray of wrist', as the views will not be adequate. Fractures of the scaphoid often elude primary detection; flexion and extension views using a curved cassette raise the frequency of primary diagnosis dramatically. Ask for these in doubtful cases.

If you intend to withhold radiography which the patient expects, make your reasons clear and persuasive. If it comes to an insoluble disagreement give in gracefully or subsequent litigation may well do more harm than another dose of X-radiation.

YAWS

A chronic relapsing non-venereal treponematosis with regional variations and nomenclature, universal throughout tropical and sub-tropical regions. May often present in emergency units with ulcerative or granulomatous forms. Recent recrudescence in many areas. Causative organism *Treponema pertenue*.

Also known as framboesia, pian, bonta, parangi, sibbens (in Scotland). Pinta in South America is due to *Treponema carateum*.

Later manifestations of the disease include osteitis, chrondritis (of nose and ears), and a wide variety of soft-tissue disorders. All respond to penicillin — penicillin G 1.2 mega in oily base given deeply IM is sufficient for primary lesions. Later manifestations need longer treatment.

The hoped-for elimination of yaws has been defeated by

poverty, ignorance, and inadequate resources for public health measures.

ZIP-FASTENER CALAMITIES

Not uncommonly unfortunates present themselves with penile skin caught in the zip of their trousers — usually a fold of foreskin which is trapped in the slide. This can be remedied by destruction of the zip mechanism with a small pair of strong wire-cutters. A little local lignocaine 1% injected at the base of the affected fold of skin makes it easier for patient and casualty officer alike. It is rare for any suturing or other surgical work to be needed afterwards.

Bibliography

Anaesthesia

The illustrated handbook in local anaesthesia, ed. Eriksson (Lloyd-Luke 1979, 2nd edn). An admirably illustrated guide to all forms of regional blockade anaesthesia.

Intensive care

Intensive care, by Emery and others (Hodder 1984, revised edn). Quite the best and easiest to understand; equally suitable for medical and nursing staff; sufficient material for anyone wanting a conspectus.

Medicine

Control of communicable disease in man, by Benenson (American Public Health Association 1975, 12th edn). Very useful; complete, brief, and explicit.

Diagnosis and management of medical emergencies, by Vakil and Udwadia (Oxford University Press 1977, 2nd edn). Systematic and excellent. No European rival is known. Its Indian weighting is no disadvantage.

Price's textbook of the practice of medicine, ed. Bodley Scott (Oxford University Press 1978, 12th edn).

Neurology

Aids to the investigation of peripheral nerve injuries, MRC memorandum No. 7 (HMSO 1975, 2nd edn). Uniquely useful and complete.

Management of head injuries, by Jennett and Teasdale (Davis 1981). Splendid book — Glaswegian provenance.

Poisons, general

The British National Formulary, ed. Brown (The Pharmaceutical Press 1982). Very good indeed; updated quarterly.

Clinical pharmacology, by Laurence (Churchill Livingstone 1980, 5th edn).

Environmental and industrial health hazards, by Trevethick (Heinemann 1976, revised edn).

Forensic toxicology, ed. Ballantyne (Wright 1974). Interesting symposium.

Poisoning: diagnosis and treatment, ed. Vale and Meredith (Update 1981). Not a complete guide, but of a high standard and very well presented.

Treatment of common acute poisonings, by Matthew and Lawson (Churchill Livingstone 1979, 4th edn). By far the best general reference book; brief, readable, and excellent, without rival; unreservedly recommended.

Recent articles and miscellaneous information, filed alphabetically by name of drug, should be kept readily available in a file of recent toxicological data.

Poisons, agricultural and commercial

Approved products for farmers and growers, issued by Ministry of Agriculture, Fisheries and Food (HMSO 1983). Published 1 year in arrears.

Clinical toxicology of commercial products, by Gosselin (Williams and Wilkins 1984, 5th edn). The only available source book; very extensive and very useful but needs practice in use because of American provenance and complicated layout; like all source books it is out of date after a year or so; many limitations, especially cost and transatlantic nomenclature.

Industrial toxicology, by Hamilton and Hardy (Wright 1983, 4th edn).

Martindale's extra pharmacopoeia, ed. Reynolds (The Pharmaceutical Press 1982, 28th edn).

Transport emergency cards (Chemical Industries Association 1980). Important source book of toxic lorry-freights. Chemsafe manual (supplied free) tells you how to get assistance. Any A & E department sited near a road carrying such chemical cargoes should have this work of reference. The publishers, Chemical Industries Association Ltd, are at Alembic House, 93 Albert Embankment, London SE1.

Poisons, naturally occurring

British poisonous plants (HMSO Reference Book 161), by Forsyth (HMSO 1980, 6th edn).

A colour atlas of poisonous plants, by Frohne and Pfänder (Wolfe Scientific 1984). Authoritative guide with good illustrations; comprehensive and reliable.

Skins

Dermatology: an illustrated guide, by Fry (Update 1978, 2nd edn).

Statistics

Epidemiology for the uninitiated, by Rose and Barker (BMA 1979).

Statistics at square one, by Swinscow (British Medical Association 1980, 6th edn). Admirably combines brevity and clarity. Anyone doing elementary research needs it.

Surgery

Cope's early diagnosis of the acute abdomen, revised by Silen (Oxford 1983, 16th edn).

Fundamental techniques of plastic surgery and their surgical applications, by McGregor (Churchill Livingstone 1980, 7th edn).

High velocity missile wounds, by Owen-Smith (Arnold 1981).

Pye's surgical handicraft, ed. Kyle (Wright 1977, 20th edn). Needs no recommendation.

Short practice of surgery, by Bailey and Love, ed. Rains and Ritchie (H. K. Lewis 1981, 18th edn).

Surgery of violence (British Medical Association 1976). Collected articles from the Royal Victoria Hospital, Belfast. Out of print.

Trauma

Accident and emergency medicine, by Rutherford and others (Pitman Medical 1980). An impressive successor to Peter London's classic tome.

Care of the acutely ill and injured (5th International Congress of Emergency Surgery), ed. Wilson and Marsden (Wiley 1982).

The casualty officer's handbook, by Wilson and Hall (Butterworth 1979, 4th edn). Worthy successor to Maurice Ellis's familiar and well-loved book.

Closed treatment of common fractures, by Chamley (Churchill Livingstone 1970, 3rd edn).

Essential accident and emergency care, by Wilson and others (MTP Press 1981). Outstanding handbook, primarily for nurses.

Gray's anatomy, ed. Warwick and Williams (Churchill Livingstone 1980, 36th British edn). The basis of the treatment of trauma.

The hand: examination and diagnosis (Churchill Livingstone 1983, 2nd edn). Brief and brilliant.

Illustrations of regional anatomy, by Jamieson, ed. Walmsley and Murphy, 7 parts (Churchill Livingstone 1971—82, 9th/10th edn). Especially useful for limb and hand injuries.

Medical aid at accidents, by Snook (Update 1974). Definitive for ambulance and fire service crews; useful for surgical flying squads.

Operative surgery: the hand, ed. Pulvertaft (Butterworth 1977, 3rd edn).

Pathology of injury, ed. Hunt (Harvey Miller and Metcalf 1972). Research-oriented and excellent.

The practical management of head injuries, by Potter (Lloyd-Luke 1974, 3rd edn). Contains the best summary of eye signs in head injuries; useful in general too.

A simple guide to trauma, by Huckstep (Churchill Livingstone 1982, 3rd edn). An eye-catching, quick reference book. Useful when under pressure.

A system of orthopaedics and fractures, by Apley (Butterworth 1982, 6th edn). Short, complete, and clear.

Topical reviews in accident surgery, ed. Tubbs and London, 2 vols. (Wright 1980, 1982).

Trauma management, by Cave and others (Year Book Medical Publishers Chicago 1979).

Trauma surgery, by Powley (Wright 1973). A very useful review of the subject; excellent on closed chest injuries.

170   Bibliography

Miscellaneous

Accident and emergency paediatrics, by Valman (Blackwell Scientific 1979, 2nd edn).

Current emergency diagnosis and treatment, by Mills, Ho, and Trunkey (Lange 1983). An impressive and informative presentation from the US 'emergency room'. Buy it and read it.

The diagnosis and primary care of accidents and emergencies in children, by Illingworth (Blackwell Scientific 1982, 2nd edn). Unique.

Ocular emergencies (Smith and Nephew Pharmaceuticals 1975). Elementary and clearly illustrated. Professionally approved.

A paediatric vade-mecum, ed. Wood (Lloyd-Luke 1982, 10th edn). A great comfort in times of doubt with simple tables of drugs, treatments, and dosages. Well indexed.

Positioning in radiography, by Clark, 2 vols (Heinemann 1979, 1981, 10th edn).

Periodicals (UK only)

Archives of Emergency Medicine
British Journal of Accident and Emergency Medicine
British Medical Journal
Hospital Medicine
Hospital Update
Injury
The Lancet
Medicine
Proceedings of the Royal Society of Medicine

Index

Main headings are shown in **BOLD CAPITALS** and subheadings in **bold type** as in the text, and are immediately followed by the principal page reference(s); other references to the same subject follow, after a semi-colon.